Mathematics, Computer Science and Logic - A Never Ending Story

Peter Paule

Editor

Mathematics, Computer Science and Logic - A Never Ending Story

The Bruno Buchberger Festschrift

 Springer

Editor
Peter Paule
Research Institute for Symbolic
   Computation
Johannes Kepler University
Linz, Austria

ISBN 978-3-319-34682-3   ISBN 978-3-319-00966-7 (eBook)
DOI 10.1007/978-3-319-00966-7
Springer Cham Heidelberg New York Dordrecht London

Springer is part of Springer Science+Business Media (www.springer.com)

Preface

Bruno Buchberger passed the milestone of his 60th birthday on October 22, 2002.

All the contributors to this book helped to celebrate this event, by presenting invited talks at the birthday conference "Logic, Mathematics and Computer Science - LMCS2002" presented in Professor Buchberger's renovated medieval castle at RISC in Hagenberg, Austria. Because of the superb spirit and the success of this symposium, the idea was launched to make these talks available to a larger audience. After more than a decade, the plan has finally come true, in the form of this collection of mathematical essays. Two of them are almost unchanged versions of the LMCS2002 talks: Stephen Wolfram's "New Directions in the Foundations of Mathematics" and Doron Zeilberger's "Towards a Symbolic Computational Philosophy (and Methodology!) for Mathematics". The essay "On the Role of Logic and Algebra in Software Engineering" by Manfred Broy is a slightly edited version of his LMCS2002 talk. Henk Barendregt significantly expanded his talk on "Foundations of Mathematics from the Perspective of Computer Verifcation". In their mathematical essence all these contributions are still fully up-to-date, and they rekindle the inspiring atmosphere of the Buchberger Symposium.

I want to take the opportunity to thank Ralf Hemmecke for editorial assistance and, last but not least, Martin Peters and Ruth Allewelt from Springer for their help and almost infinite patience.

Hagenberg, Austria   Peter Paule
May 2013

Contents

Foundations of Mathematics
from the Perspective of Computer Verification

Henk Barendregt

To Bruno Buchberger independently of any birthday

Abstract In the philosophy of mathematics one speaks about Formalism, Logicism, Platonism and Intuitionism. Actually one should add also Calculism. These foundational views can be given a clear technological meaning in the context of Computer Mathematics, that has as aim to represent and manipulate arbitrary mathematical notions on a computer. We argue that most philosophical views over-emphasize a particular aspect of the mathematical endeavor.

1 Mathematics

The ongoing creation of mathematics, that started 5 or 6 millennia ago and is still continuing at present, may be described as follows. By looking around and abstracting from the nature of objects and the size of shapes *homo sapiens* created the subjects of arithmetic and geometry. Higher mathematics later arose as a tower of theories above these two, in order to solve questions at the basis. It turned out that these more advanced theories often are able to model part of reality and have applications. By virtue of the quantitative, and even more qualitative, expressive force of mathematics, every science needs this discipline. This is the case in order to formulate statements, but also to correct conclusions (Fig. 1).

H. Barendregt (✉)
Nijmegen University, Nijmegen, The Netherlands
e-mail: henk@cs.ru.nl

P. Paule (ed.), *Mathematics, Computer Science and Logic - A Never Ending Story*,
DOI 10.1007/978-3-319-00966-7_1, © Springer International Publishing Switzerland 2013

Fig. 1 The triangle of
mathematical activities

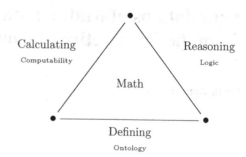

The mathematical endeavor consists in a stylized way of three activities[1]: *defining, calculating* and *proving.*[2] The three started in this order, but over the centuries they became more and more intertwined. Indeed, before one can do arithmetic, one has to have numbers and an analogous statement holds for geometry. Having numbers one wants to add and multiply these; having polygons one wants to calculate their area. At some point the calculations became complex and one discovered shortcuts. One role of proofs is that they are an essential tool to establish the correctness of calculations and constructions.

1.1 Egyptian-Chinese-Babylonian vs Greek Mathematics

Different appreciations of the three sides of the triangle of mathematical activities gave rise to various explicit foundational views. Before entering into these we will argue that different implicit emphases on the sides of the triangle also did lead to different forms of mathematics. In the Egyptian-Chinese-Babylonian tradition emphasis was put on calculation. One could solve e.g. linear and quadratic equations. This was done in a correct way, but a developed notion of proof was lacking. In the Greek tradition the emphasis was on proofs. Using these one can show that there are infinitely many primes, or that $\sqrt{2}$ is irrational, something impossible to do by mere computation alone. But the rigor coming from geometric proofs also had its limitations. Euclid[3] [51] gives a geometric proof that $(x+y)^2 = x^2 + 2xy + y^2$, but no similar results for $(x+y)^3$ (although such a result could have been proved geometrically) or $(x+y)^4$, let alone $(x+y)^n$.

---

[1] I learned this from Gilles Barthe (1996, personal communication).

[2] The activity of *solving* can be seen as a particular instance of computing (or of proving, namely that of an existential statement $\exists x.P(x)$ in a constructive setting).

[3] App. 325–265 BC.

Then came Archimedes (287–212 BC), who was well versed in both calculating and proving. Another person developing mathematics toward the synthesis of these two traditions was the Persian mathematician al-Khowârizmî (app. 780–850 AD), who showed that the algorithms for addition and multiplication of decimal numbers (as we learn them at school) are provably correct.

When calculus was invented by Newton (1643–1727) and Leibniz (1646–1716) the dichotomy between proving and computing was reinforced. Newton derived Kepler's laws of planetary movement from his own law of gravitation. For this he had to develop calculus and use it in a nontrivial way. He wanted to convince others of the correctness of what he did, and went in his Principia into great detail to arrive at his conclusions geometrically, i.e. on the Greek tradition.[4] Leibniz [83] on the other hand used calculus with a focus on computations. For this he invented the infinitesimals, whose foundation was not entirely clear. But the method worked so well that this tradition still persists in physics textbooks. Euler could do marvelous things with this computational version of calculus, but he needed to use his good intuitio in order to avoid contradictions. Mathematicians in Britain, on the other hand, "did fall behind" by the Greek approach of Newton, as stated by Kline (1908–1992) [77], pp. 380–381. Only in the nineteenth century, by the work of

---

[4]Newton also did many important things for the synthesis of the two styles of doing mathematics. His binomial formula $(x + y)^n = \sum_{k=0}^n \binom{n}{k} x^{n-k} y^k$ involves computing and reasoning. It also makes sense for n a rational number. Also his fast method of computing digits of π, see [96] or [21] pp. 142–143, is impressive. By computing twice

$$\int_0^{\frac{1}{4}} \sqrt{x - x^2} dx,$$

one time using calculus, another time using planar geometry and employing the binomial formula for $n = \frac{1}{2}$, Newton derived

$$\pi = 24\left(\frac{\sqrt{3}}{32} + \frac{1}{12} - \frac{1}{160} - \frac{1}{3,584} - \frac{1}{36,864} - \frac{5}{1,441,792} - \frac{7}{13,631,488} \cdots\right)$$

$$= 24\left(\frac{\sqrt{3}}{32} + \frac{1}{3}\frac{1}{2^2} - \frac{1}{5}\frac{1}{2^5} - \frac{1}{7}\frac{1}{2^9} - \frac{1}{9}\frac{1}{2^{12}} - \sum_{k=4}^{\infty} \frac{2k-3}{(2k+1)2^{3k+5}}\right),$$

using modern notation. Newton knew how to compute $\sqrt{3}$ and this series converges quite fast. In this way he obtained $\pi = 3.14159265897928$, the last two digits are a roundoff error for 32. Ludolph van Ceulen (1539–1610) spent several decades of his life in order to compute 32 digits (later 35 digits published on his tomb in Leiden), see his [119], while with Newton's method this could have been done in a day or so. As opposed to Newton it should be admitted that van Ceulen was more precise about the validity of the digits he obtained.

Cauchy (1789–1857) and Weierstrass[5] (1815–1897), the computational and proving styles of doing calculus were unified and mathematics flourished as never before.[6]

In the last third of the twentieth century the schism between computing and proving reappeared. Computer Algebra Systems are good at symbolic computing, but they cannot keep track of assumptions and use them to check whether the side conditions necessary for certain computations actually hold, nor provide proofs of the correctness of their results. Proof-verification Systems at first were not good at computing and at providing proofs for the correctness of the result of a computation. This situation is changing now.

1.2 Progress on Foundations

During the development of mathematics, notations have been introduced to help the mathematicians to remember what they defined and how, and what they did compute and prove. A particularly useful notation came from Vieta (1540–1603), who introduced variables to denote arbitrary quantities. Together with the usual notations for the algebraic operations of addition and multiplication, this made finding solutions to numerical problems easier. The force of calculus consists for a good part in the possibility that functions can be manipulated in a symbolic way.

During the last 150 years general formal systems have been introduced for defining, computing and reasoning. These are the formal systems for ontology, computability and logic. The mathematical notations that had been used throughout the centuries now obtained a formal status. If a student who states the Archimedian axiom as "For all x and all $\epsilon>0$ there exists an $n \in \mathbb{N}$ such that $n\epsilon$ is bigger" a teacher could say only something like: "I do not exactly understand you." If the student is asked to use a formal statement to express what he or she means and answers "$\forall x \forall \epsilon>0 \, \exists n \in \mathbb{N}.n\epsilon>$" the teacher can now say that this is demonstrably not a WFF (well formed formula). This little example is enough to show that these

---

[5]Poincaré (1854–1912) made a distinction between logicians using "Analysis", among which he placed Weierstrass, and intuitive mathematicians, using "Synthesis", like Klein. He mentioned that the intuitive mathematicians are better in discovery, although some logicians have this capacity as well. Poincaré added that we need both types of mathematicians: *Les deux sortes d'esprits sont également nécessaires aux progrès de la science; les logiciens, comme les intuitifs, ont fait de grandes choses que les autres n'auraient pas pu faire. Qui oserait dire s'il aimerait mieux que Weierstrass n'eût jamais écrit, ou s'il préférerait qu'il n'y eût pas eu de Riemann?* See [102], Chap. 1: L'intuition et la logique en mathématiques.

[6]In the nineteenth century the infinitesimals of Leibniz were abolished (at least in mainstream mathematics). But in the twentieth century they came back as *non-standard* reals. One way of doing this is by considering $h > 0$ as infinitesimal if $\forall n \in \mathbb{N}.h<\frac{1}{n}$; for this it is necessary to work in a non-Archimedian extension of $\mathbb{R}$, which can be obtained as $\mathbb{R}^I/D$, where I is an infinite set and D is an ultra-filter on $\mathcal{P}(I)$. This approach is due to Robinson (1918–1974), see his [105]. The other way consist of infinitesimals $h > 0$, such that $h^2 = 0$. This time the trick is to work in an intuitionistic context where the implication $h^2 = 0 \Rightarrow h = 0$ does not hold, see [94] and [23].

systems do provide help with defining. They also provide help with computing and proving. Often it is felt by working mathematicians, that formalization acts as a kind of chastity belt.[7] There are, nevertheless, good reasons for formalization. It may provide extra clarity to determine whether a certain reasoning is correct. But there is more to it: formalizations allow a strong form of *reflection* over established mathematics, to be discussed below.

It was mentioned that there are mathematical disciplines dealing with one subject, like arithmetic and geometry, and that there are "towers of theories" above these subjects. Examples of the latter are algebraic and analytical number theory. Let us call mathematics of the first kind a *close-up* and of the second kind a *wide-angle* discipline. These two styles will have their counterparts in the foundations of mathematics.

1.3 Reflection

There is one aspect of mathematics, present in informal mathematics as well as formalized mathematics, that plays an important role. This is *reflection* over the syntactic form of the mathematics that has been obtained so far. An early example of reflection in informal mathematics is the duality principle in planar projective geometry discovered by Poncelet (1788–1867), Gergonne (1771–1859) and Plücker (1801–1868): "Given a theorem one obtains another one by replacing the word point by line and vice versa.[8]"

Another example comes from Mostowski[9](1968, personal communication), who challenged in the 1960s an automated theorem prover by asking whether

$$(((((((((((((((A \leftrightarrow A) \leftrightarrow A) \leftrightarrow A) \leftrightarrow A) \leftrightarrow A) \leftrightarrow A) \leftrightarrow A) \leftrightarrow A) \leftrightarrow A)$$
$$\leftrightarrow A) \leftrightarrow A) \leftrightarrow A) \leftrightarrow A) \leftrightarrow A) \leftrightarrow A) \leftrightarrow A) \leftrightarrow A) \leftrightarrow A) \leftrightarrow A$$

is a propositional tautology. The machine did not give an answer, but the statement can be seen as a tautology by considering it as A_{20} with

$$A_1 = A$$
$$A_{n+1} = A_n \leftrightarrow A$$

for which sequence one can show by induction on n that A_{2n} holds for all $n \in \mathbb{N}^+$.

---

[7]On the other hand, some constructivists may call some formal systems, notably those for classical set theory, a license for promiscuity.

[8]It is convenient to first replace the statements '*point P lies on line l*' and '*line l goes through point P*' by the statements '*P touches l*' and '*l touches P*', respectively.

[9]1913–1975.

Using formal systems also allows a form of reflection on the mathematical activities. This gives metamathematical results, showing e.g. the limitations of the methods involved (incompleteness and undecidability theorems). The incompleteness theorem of Gödel[10] [62] uses reflection in a fundamental way: provability becomes internalized. Not only this famous theorem, but also the completeness theorem

$$\Gamma \vdash A \iff \Gamma \models A$$

(a statement A is derivable from the axioms Γ in first order logic iff A holds in all structures in which the axioms Γ are valid). Most other metamathematical results use reflection: within mathematics one speaks about mathematical statements.

Besides pointing to limitations, reflection also enables us to "get more mileage". If we know that a result can be proved using first order logic, then by the compactness theorem we sometimes can come to stronger conclusions. In fact this is the starting point of model theory. This subject was at first mainly based on reflection over the mathematical activities of proving and defining. Later notions like "computably (recursively) saturated models", see Chang and Keisler [33], showed that model theory fruitfully makes use of reflection over the full triangle of mathematical activities.

Also set theory and category theory know their forms of reflection. In set theory there is Gödel's construction of the class L of constructible sets, which uses internalized definability. In category theory one can internalize the notion of a category inside a topos.

2   Foundational Formalisms

In this section we will describe various formalisms for the mathematical endeavor in the following order: logic, computability and ontology.

2.1   Logic

The Greek philosopher Aristotle (384–322 BC) made several fundamental contributions to the foundations of mathematics that are still relevant today. From him we have inherited the idea of the *axiomatic method*,[11] not just for mathematics, but for all sciences. A science consists of statements about concepts. Concepts have to be *defined* from simpler concepts. In order to prevent an infinite regression, this process starts from the *primitive concepts*, that do not get a definition. Statements have to be

---

[10]1906–1978.

[11]In [7], Posterior Analytics.

proved from statements obtained before. Again one has to start somewhere; this time the primitive statements are called *axioms*. A statement derived from the axioms by pure reason is called a theorem in that axiomatic system. In mathematics one starts from arbitrary primitive notions and axioms, while in science from empirical observations, possibly using (in addition to pure reason) the principle of induction (generalization).

Just a couple of decades after Aristotle and the axiomatic method, Euclid came with his compilation of existing geometry in this form in his *Elements*[12] and was very influential as an example of the use of the axiomatic method. Commentators of Euclid stated that the primitive notions are so clear that they did not need definitions; similarly it was said that the axioms are so true that they did not need a proof. This, of course, is somewhat unsatisfactory. A couple of millennia later Hilbert (1862–1943) changed this view. For him it did not matter what exactly is the essence of the primitive notions such as point and line, as long as they satisfy the axioms: *"The axioms form an implicit definition of the primitive concepts"*. This is a fair statement, even if not all axiom systems determine up to isomorphism the domains of concepts about which they speak. In fact the more traditional mathematical theories (arithmetic, geometry) may have had the intention to describe precisely a domain thought to be unique.[13] In modern mathematics, axioms are often used with almost the opposite intention: to capture what it is that different structures have in common. The axioms of group theory[14] describe groups of which there are many kinds and sizes.

It was again Aristotle who started the quest for logic, i.e. the laws by which scientific reasoning is possible.[15] Aristotle came up with some *syllogisms* (valid reasoning step based on syntactical form) like

$$\frac{\text{No } A \text{ is a } B}{\text{No } B \text{ is a } A}.$$

Aristotle explains this by the particular case

$$\frac{\text{No horse is a man}}{\text{No man is a horse}}.$$

---

[12]As was already observed in antiquity the theorems in the Elements were not always proved from the axioms by logic alone. Sometimes his arguments required extra assumptions. The axiomatization of [68] corrected the subtle flaws in Euclid.

[13]By Gödel's incompleteness theorem the axioms of arithmetic do not uniquely determine the set of natural numbers. By the existence of different forms of geometry and its applicability in physics we know that Euclidean geometry is not the only variant and not even the true theory about space.

[14]Actually this well-known axiom system (of a set with a binary operation such that $\forall x, y \, \exists z \, x \cdot z = y$) is a close-up theory of statements that are valid in arbitrary groups. Next to this there is also the much more interesting wide-angle theory of groups studied with their interconnections and occurrences in mathematical situations.

[15]In [7], Prior Analysis. One may wonder whether his teacher Plato (427–347 BC) was in favor of this quest (because we already *know* how to reason correctly).

Another of his syllogisms is

$$\frac{\text{No } A \text{ is a } C \quad \text{All } B \text{ are } C}{\text{No } A \text{ is a } B}.$$

Take e.g. men, swans and birds for A, B and C respectively. Aristotle also makes a distinction between such syllogisms and so called *imperfect* syllogisms, that require more steps (nowadays these are called admissible rules). The idea of specifying formal rules sufficient for scientific reasoning was quite daring and remarkable at the time. Nevertheless, from a modern perspective the syllogisms of Aristotle have the following shortcomings. (1) Only unary predicates are used (monadic logic). (2) Only composed statements involving $\rightarrow$ and $\forall$ are considered (so $\&$, $\vee$, $\neg$ and $\exists$ are missing). (3) The syllogisms are not sufficient to cover all intuitively correct steps.

In commentators of Aristotle one often finds the following example.

$$\frac{\text{All men are mortal} \quad \text{Socrates is a man}}{\text{Socrates is mortal}}. \tag{1}$$

Such 'syllogisms' are not to be found in Aristotle, but became part of the traditional logical teaching. They have an extra disadvantage, as they seem to imply that they do need to lead from true sentences to true sentences. This is not the case. Syllogism only need to be truth preserving, even if that truth is hypothetical. So a more didactic (and more optimistic) version of (1) is

$$\frac{\text{All sentient beings are happy} \quad \text{Socrates is a sentient being}}{\text{Socrates is happy}}. \tag{2}$$

This example is more didactic, because one of the premises is not true, while the rule is still valid. Aristotle was actually well aware of this hypothetical reasoning.

It was more than 2,300 years later that Frege[16] [54] completed in 1879 the quest for logic and formulated (first-order) predicate logic. That it was sufficient for the development of mathematics from the axioms was proved by Skolem[17] [110] and independently by Gödel [61] as the completeness theorem for first order logic; see [74] for the priority of Skolem.

The final fundamental contribution of Aristotle to modern foundations of mathematics is his distinction between proof-checking and theorem-proving. He remarked that if someone would claim to have a proof of a statement and present it, then that proof always could be checked line by line for its correctness. If someone would claim that a statement is provable, without presenting the proof, then it is much harder to verify the correctness of this assertion. From a modern perspective these

---

[16]1848–1925.
[17]1887–1963.

Elimination rule	Introduction rule
$\dfrac{\Gamma \vdash A \quad \Gamma \vdash A \to B}{\Gamma \vdash B}$	$\dfrac{\Gamma, A \vdash B}{\Gamma \vdash A \to B}$
$\dfrac{\Gamma \vdash A \,\&\, B \quad \Gamma \vdash A \,\&\, B}{\Gamma \vdash A \qquad \Gamma \vdash B}$	$\dfrac{\Gamma \vdash A \quad \Gamma \vdash B}{\Gamma \vdash A \,\&\, B}$
$\dfrac{\Gamma \vdash A \vee B \quad \Gamma, A \vdash C \quad \Gamma, B \vdash C}{\Gamma \vdash C}$	$\dfrac{\Gamma \vdash A \qquad \Gamma \vdash B}{\Gamma \vdash A \vee B \quad \Gamma \vdash A \vee B}$
$\dfrac{\Gamma \vdash \forall x.A}{\Gamma \vdash A[x := t]}\ t \text{ is free in } A$	$\dfrac{\Gamma \vdash A}{\Gamma \vdash \forall x.A}\ x \notin \Gamma$
$\dfrac{\Gamma \vdash \exists x.A \quad \Gamma, A \vdash C}{\Gamma \vdash C}\ x \notin C$	$\dfrac{\Gamma \vdash A[x := t]}{\Gamma \vdash \exists x.A}$

Start rule		False rule	Double-negation rule
$\dfrac{}{\Gamma \vdash A}\ A \in \Gamma$		$\dfrac{\Gamma \vdash \perp}{\Gamma \vdash A}$	$\dfrac{\Gamma \vdash \neg\neg A}{\Gamma \vdash A}$

Fig. 2 Predicate logic natural deduction style

remarks may be restated as follows: proof-checking is decidable, while theorem testing is (in general) impossible, i.e. undecidable.

2.1.1 Rules of Logic

The rules of logic as found by Frege have been given a particularly elegant form by Gentzen[18] [58], in his system of natural deduction, see Fig. 2. Predicate logic gives in the first place a basis for close-up mathematics (as defined in Sect. 1). Using the axioms of (Peano) arithmetic or Euclidean geometry (as provided rigorously by Hilbert) together with the deductive power of logic one can prove a good deal. For wide-angle mathematics one needs a stronger formal system. Below we will meet three candidates for this: set theory or category theory combined with logic, or type theory (in which a sufficient amount of logic is built in).

Several comments are in order. Γ stands for a set of formulas and $\Gamma, A = \Gamma \cup \{A\}$. The formula $\perp$ stands for absurdum, i.e. the false statement. Negation is defined as

---

[18] 1909–1945.

Fig. 3 The primitive
recursive functions

$$
\begin{array}{rcl}
Z(x) &=& 0; \\
S(x) &=& x+1; \\
P_k^n(x_1,\dots,x_n) &=& x_k; \\
\hline
f(\vec{x}) &=& g(h_1(\vec{x}),\dots,h_n(\vec{x})); \\
\hline
f(\vec{x},0) &=& g(\vec{x}); \\
f(\vec{x},y+1) &=& h(\vec{x},y,f(\vec{x},y)).
\end{array}
$$

$\neg A = (A \rightarrow \bot)$. The double negation rule is left out for Intuitionistic logic to be discussed later. The condition $x \notin FV(C)$ ($x \notin FV(\Gamma)$) means that x is not among the free variables in C (in a formula of Γ respectively).

2.2 Computability

A theoretical foundation for the act of calculating was given relatively late. Following work of Grassmann (1809–1877), Dedekind (1831–1916), Peano (1858–1932) and Skolem it was Hilbert [69] who formally introduced a class of schematically defined numerical functions (i.e. on the set $\mathbb{N}$ of natural numbers). In Gödel [62] this class was called the *recursive* functions but, he was aware that it did not encompass all computable functions (by a human calculator). Their usual name *primitive recursive functions* was given later by Rózsa Péter[19] [99]. This class can be defined as follows (Fig. 3).

Here the variables like x range over the natural numbers $\mathbb{N}$. The scheme states that Z, S and the P_k^n are (primitive) recursive functions and that and if g, h denote earlier obtained (primitive) recursive functions then so is f defined in terms of these.

2.2.1 General Computability

Sudan[20] [112] and Ackermann[21] [1] independently defined a computable function that was not primitive recursive. A simplification of such a so-called Ackermann function, with the property of not being primitive recursive but nevertheless computable, was given by Péter [100].

$$
\psi(0, y) = y + 1;
$$
$$
\psi(x + 1, 0) = \psi(x, 1);
$$
$$
\psi(x + 1, y + 1) = \psi(x, \psi(x + 1, y)).
$$

---

[19] 1905–1977.
[20] 1889–1977.
[21] 1896–1962.

$$\text{Define } f : \mathbb{N} \rightarrow \mathbb{N} \text{ by} \quad f(n) \;=\; \begin{cases} \frac{1}{2}n, & \text{if } n \text{ is even;} \\ 3n+1, & \text{else.} \end{cases}$$

$$\text{Does one have} \qquad \forall n \geq 1 \exists k. f^k(n) = 1?$$

Fig. 4 Collatz' problem

Gödel (based on an idea of Herbrand (1908–1931)) [62], Church[22] [35] and Turing[23] [116] independently introduced richer computational mechanisms: respectively systems of equations, lambda calculus and Turing machines. Kleene[24] [75] and Turing [117] proved that these systems all represent the same class of numerical functions. This class of functions computable by either of these formalisms is now generally considered as the class of humanly computable functions. Besides this, both Church [36] and [117] indicated how to specify a non-computable function. This showed that an old ideal of Leibniz could not be fulfilled. Leibniz wanted (1) to formulate a language in which all problems could be stated; (2) to construct a machine that could decide the validity of such statements.[25] Turing showed in fact that the problem to determine whether a logical formula A is derivable from some axioms Γ, (i.e. whether $\Gamma \vdash A$ holds) is undecidable.

2.2.2 Term Rewrite Systems

The notion of Herbrand-Gödel computability generalizes the class of primitive recursive functions and yields the (partial) computable functions. The idea is presented in the format of Term Rewriting Systems (TRSs) in Klop et al. [78]. Rather than presenting the precise definitions we give a representative example. The following is called Collatz' problem (Fig. 4).[26]

The following TRS describes this problem. It uses the following constants and function symbols

$$\{0, S, \text{true}, \text{false}, \text{even}, \text{odd}, \text{if}, \text{half}, \text{threeplus}, \text{syracuse}, \text{perpetual}\}.$$

---

[22]1903–1995.

[23]1912–1954

[24]1909–1994.

[25]It is said that the existence of God was the first question about which Leibniz wanted to consult his machine. This is an early example of a striking confidence in high technology.

[26]The problem is still open, and a prize of 1,000 UK pounds is offered for its solution. For $1 \leq n \leq 10^{17}$ the conjecture has been verified by Oliveira e Silva using in total 14 CPU years on 4 computers (average 200 MHz), see Chamberland (2003) *An update on the* $3x + 1$ *problem* ⟨www.math.grinnell.edu/~chamberl/papers/survey.ps⟩

$$
\begin{array}{lll}
\texttt{even(0)} & \rightarrow & \texttt{true} \\
\texttt{even(S}(x)) & \rightarrow & \texttt{odd}(x) \\
\texttt{odd(0)} & \rightarrow & \texttt{false} \\
\texttt{odd(S}(x)) & \rightarrow & \texttt{even}(x) \\
\texttt{threeplus(0)} & \rightarrow & \texttt{S(0)} \\
\texttt{threeplus(S}(x)) & \rightarrow & \texttt{S(S(S(threeplus(x)))))} \\
\texttt{if(true},x,y) & \rightarrow & x \\
\texttt{if(false},x,y) & \rightarrow & y \quad \% \ \texttt{if}(b,x,y) = \texttt{if } b \texttt{ then } x \texttt{ else } y \\
\texttt{syracuse}(x) & \rightarrow & \texttt{if(even}(x),\texttt{half}(x),\texttt{threeplus}(x))
\end{array}
$$

$$
\% \ \texttt{syracuse}(n) = \begin{cases} \texttt{half}(n) & \text{if } \texttt{even}(n) \\ \texttt{threeplus}(n) & \text{else} \end{cases}
$$

$$
\begin{array}{lll}
\texttt{perpetual(0)} & \rightarrow & \texttt{0} \\
\texttt{perpetual(S(0))} & \rightarrow & \texttt{S(0)} \\
\texttt{perpetual(S(S}(x))) & \rightarrow & \texttt{perpetual(syracuse(S(S}(x))))
\end{array}
$$

$$
\% \text{Conjecture: } \forall n \in \mathbb{N}^+.\texttt{perpetual}(\underline{n}) \twoheadrightarrow 1\mathord{??}
$$

Fig. 5 A TRS for Collatz' problem

We write $\underline{0} = 0$, $\underline{n+1} = S(\underline{n})$ and add comments for the intended meanings (Fig. 5).

Note that there are no rewrite rules with $0, S, \texttt{true}, \texttt{false}$ at the 'head' of a LHS (left hand side): these are the so called *constructors*. The other elements of the alphabet are (auxiliary) function symbols. One may reduce (use the rewrite rules $\rightarrow$) within terms at arbitrary positions of the *redexes*, i.e. left hand sides with for the variables (x, y) substituted arbitrary terms. One easily shows that in this TRS ($\twoheadrightarrow$ denotes many step reduction)

$$
\texttt{perpetual}(\underline{n}) \twoheadrightarrow 1 \ \Leftrightarrow \ \exists k \in \mathbb{N}.\, f^k(n) = 1.
$$

The Collatz conjecture now becomes $\forall n {>} 0.\ \texttt{perpetual}(\underline{n}) \twoheadrightarrow 1$. Erdös remarked: "Mathematics is not yet ready for these kinds of problems", referring to the Collatz conjecture. This fashion of defining computable partial functions via Term Rewrite Systems is now the standard in functional programming languages, such as Clean[27] and Haskell.[28]

2.2.3 Combinatory Logic

As usual when dealing with universal computability, there is in fact a universal mechanism: a single TRS in which all computable functions can be represented. This is **CL**, combinatory logic, due to Schönfinkel (1889–1942?) and Curry (1900–1982), see Klop et al. [78]. This TRS has two constants **S,K** and a binary function symbol app (Fig. 6).

---

[27] ⟨www.cs.kun.nl/~clean⟩

[28] ⟨www.haskell.org⟩

$$\begin{array}{ll}
\mathrm{app}(\mathrm{app}(\mathrm{app}(\mathsf{S}, x), y), x) & \rightarrow \quad \mathrm{app}(\mathrm{app}(x, z), \mathrm{app}(y, z)) \\
\mathrm{app}(\mathrm{app}(\mathsf{K}, x), y) & \rightarrow \quad x
\end{array}$$

Fig. 6 CL, functional notation

$$\begin{array}{ll|ll}
\mathsf{S} \cdot x \cdot y \cdot x & \rightarrow \quad x \cdot z \cdot (y \cdot z) & \mathsf{S}xyx & \rightarrow \quad xz(yz) \\
\mathsf{K} \cdot x \cdot y & \rightarrow \quad x & \mathsf{K}xy & \rightarrow \quad x
\end{array}$$

Fig. 7 CL infix notation and applicative notation

It is more usual to give this TRS in an equivalent different notation (Fig. 7).

The price for being universal is non-termination. Indeed, it is easy to give a non-terminating CL expression: (SII)(SII), where I = SKK. A more interesting one is S(SS)SSSS.[29] Potentially non-terminating TRSs are important if a proof-search is involved.[30]

2.3 Ontology

The branch of philosophy that deals with existence is called *ontology*. Kant stated that 'being' is not a predicate. Indeed, if we would state $B(x)$ with as intended meaning that x exists, we already need to assume that we have the x whose existence is asserted. But in axiomatic theories the notion of 'being' makes sense. One has formal expressions and one claims for only some of these that they make sense, i.e. exist. For example $\frac{1}{0}$ and $\{x \mid x \notin x\}$ are expressions in respectively arithmetic and set theory whose existence cannot be consistently asserted.

In mathematics before roughly 1800 only less than a dozen basic domains are needed: the number systems $\mathbb{N}$, $\mathbb{Z}$, $\mathbb{Q}$, $\mathbb{R}$, $\mathbb{C}$, the Euclidean and projective plane and space and perhaps a few more. Since Descartes, one used the construction of products of domains. But that was more or less all. In the nineteenth century a wealth of new spaces were created that, as stated in Sect. 1, have their impact for proofs of properties of the elements in the more familiar spaces. Examples are groups, non-Euclidean spaces, Riemann surfaces, function spaces and so on. Therefore the need for a systematic ontology arose.

---

[29]Due to M. Baron and M. Duboué, see [10], Exercise 7.4.5(i).

[30]In [59] it is proved that one can safely add the fixed-point combinators Y (one for each type) satisfying

$$\mathsf{Y} f \twoheadrightarrow f(\mathsf{Y} f)$$

to the proof-assistant Coq. 'Safely' means that if a putative proof containing Y normalizes, than that normal form is a proof in the system without the Y.

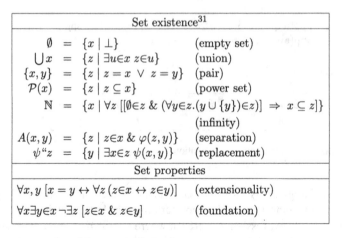

Set existence[31]		
$\emptyset = \{x \mid \bot\}$		(empty set)
$\bigcup x = \{z \mid \exists u{\in}x \; z{\in}u\}$		(union)
$\{x,y\} = \{z \mid z = x \lor z = y\}$		(pair)
$\mathcal{P}(x) = \{z \mid z \subseteq x\}$		(power set)
$\mathbb{N} = \{x \mid \forall z\,[[\emptyset{\in}z \;\&\; (\forall y{\in}z.(y \cup \{y\}){\in}z)] \Rightarrow x \subseteq z]\}$		
		(infinity)
$A(x,y) = \{z \mid z{\in}x \;\&\; \varphi(z,y)\}$		(separation)
$\psi\text{``}z = \{y \mid \exists x{\in}z \; \psi(x,y)\}$		(replacement)
Set properties		
$\forall x,y \; [x = y \leftrightarrow \forall z \,(z{\in}x \leftrightarrow z{\in}y)]$		(extensionality)
$\forall x \exists y{\in}x \; \neg\exists z \,[z{\in}x \;\&\; z{\in}y]$		(foundation)

Fig. 8 A modern version of the axioms of **ZF** set theory.

2.3.1 Set Theory

The first systematic ontology for mathematics was created by Cantor (1845–1918) in the form of set theory. This (informal) theory, contained an inconsistency, as discovered by Russell (1872–1970), when it was embedded in a formal theory by Frege: the 'set' $R = \{x \mid x \notin x\}$ satisfies $R \in R \leftrightarrow R \notin R$. Zermelo (1871–1953) removed the possibility of this proof of inconsistency and the theory was later extended by Fraenkel (1891–1965) resulting in **ZF** set theory. It is a theory formulated in first order logic with equality with $\in$ and $=$ as only predicates. The axioms claim the existence of the sets $\emptyset$ and $\mathbb{N}$ and that other sets are constructed from given sets (Fig. 8) and that they satisfy some basic properties (a set is fully determined by its elements; there are no infinite decending chains $x_0 \ni x_1 \ni x_2 \ni \ldots$, where '$\ni$' stands for the inverted '$\in$' (epsilon, set membership). It doesn't work in the environment I used, but it is a standard symbol.

This theory is powerful, in the sense that it gives a place to all needed domains in modern mathematics,[32] but in some sense it is too powerful. One can form huge sets like

$$\mathcal{P}\left(\bigcup_{n \in \mathbb{N}} \mathcal{P}^n(\mathbb{N})\right),$$

---

[31] In the list of axioms of set theory, φ is a predicate on sets and ψ is a predicate that is a 'class function', i.e. one can prove $\forall x \exists! y. \psi(x,y)$. Here $\exists!$ means unique existence, making the intended meaning of $\psi\text{``}z = \text{Range}(\psi \restriction z)$. Moreover, $x \subseteq y \Leftrightarrow \forall z[z \in x \Leftrightarrow z \in y]$, $x \cup y = \bigcup\{x,y\}$ and $\{x\} = \{x,x\}$. The precise formulation of the separation and replacement axioms using first order definable formulas φ and ψ is due to Skolem.

[32] An exception is the category of all small categories (that are sets).

and much bigger. Pondering about such large sets one may feel a *"horror infiniti"*.[33] This term was coined by Cantor [31] referring to Gauss (1777–1855) who held that the infinite is only a way of speaking.[34] In set theory with the generalized continuum hypothesis this monster looks tame in a deceptive way, as it has as cardinality 'only' $\aleph_{\omega+1}$. Perhaps it was Cantor's liking of neo-Thomistic thinking that made him comfortable with the infinite.

We will discuss how a notion like the (generalized) Cartesian product can be encoded in set theory. If $x \mapsto B_x$ is a class-function that assigns to a set x a unique set B_x, then

$$\Pi_{x \in A} B_x = \left\{ f \in A \to \bigcup \{ B_x \mid x \in A \} \mid \forall x \in A. f(x) \in B_x \right\}.$$

One needs the union, pair, power set, separation and replacement axioms. (Just the ordinary Cartesian product

$$A \times B = \{\{\{a,a\},\{a,b\}\} \mid a \in A, \, b \in B\}$$

already needs several times the pair and replacement axioms.) In type theory the formation of the Cartesian product is just one axiom (but other notions, easy to formulate in set theory, will be more involved).

Often the axiom of choice is added to ZF. It is equivalent with

$$\forall x \in I. B_x \neq \emptyset \;\Rightarrow\; \Pi_{x \in I} B_x \neq \emptyset.$$

Also the existence of large cardinals is often assumed, among other things in order to ensure that there exists a category of all small categories.

2.3.2 Type Theory

Type theory provides an ontology less permissive than set theory. It should be emphasized that there are several non-equivalent versions of type theory. Intensional, extensional; first-, second- and higher-order (the second and higher-order ones often called *impredicative*); intuitionistic and classical; with none, some and full computational power; with freely generated data-types and induction/recursion principles over these (See [6, 11, 40, 89, 93] for various forms of type theory.). This

---

[33]"Horror for the infinite." Actually people have this existential experience already just thinking about $\mathbb{N}$ as a completed totality. Aristotle rejected the actual infinite, but did recognize the potential infinite. The Platonist view of mathematics likes to consider infinite sets as an actuality from which arbitrary subsets can be taken.

[34]*"But concerning your proof, I protest above all against the use of an infinite quantity as a completed one, which in mathematics is never allowed. The infinite is only a* façon de parler, *in which one properly speaks of limits".* See [57].

multitude should be seen as an advantage: by choosing a particular type theory one is able to provide a foundation for particular proofs in mathematics, e.g. first- or second-order arithmetic. It is known that one can prove more in the latter, but at the price of having to believe in the quantifying over predicates over the natural numbers. This in contrast to **ZF** set theory, where one buys all the features in one package. A type A is like a set as it may have inhabitants; if a is an inhabitant of A one writes $a : A$. Of course the transition from the notation and terminology in set theory (if $a \in A$, then a is an element of A) is just conventional. But there is more to it. Type theory is usually *intensional*, i.e. such that $a : A$ is decidable. For this reason there is no separation axiom stating that

$$A' = \{a{:}A \mid P(a)\}$$

is another type. Indeed, the predicate P may be undecidable and this would entail the undecidability of $a{:}A'$. For similar reasons there is no quotient type $A/\sim$, where $\sim$ is an equivalence relation on A. Indeed,

$$a : [b],$$

where $[b]$ is the equivalence class of b, which means that $a \sim b$ and this could be undecidable. In spite of this, it is possible to represent 'subtypes' and 'quotient types' in type theory, but then proofs come to play a role. Intuitively,

$$A' = \{(a, p) \mid a{:}A \ \& \ p \text{ is a proof of } P(a)\};$$

$$A/\sim \ = (A, \sim) \text{ considered as structure with a different equality.}$$

Extensional type theory on the other hand does not have the a priori requirement that $a : A$ is decidable. In this version of type theories one can form arbitrary sub- and quotient types. But then one needs to consider triples (p, a, A), where p is a proof of $a : A$. The extra effort is similar to the simulation of subtypes and quotient types using proofs as was discussed above.

Another feature of type theory is that a statement like $a : A$ can be interpreted in more than one way. The first was already discussed: a is an inhabitant of A. The second interpretation is that now A is considered as a proposition and a is a proof of A (*propositions-as-types* interpretation). In this way proofs become explicit whereas in ordinary logical systems they remain implicit. Indeed, in logic one writes $\vdash_L A$, meaning that A is provable in the logic L. In a type theory T one has to furnish a complete expression p such that $\vdash_T p : A$. Note that this is related to the decidability of p being a proof of A and the (in general) undecidability of provability in a theory L, as was foreseen by Aristotle. For many logics L one has a corresponding type theory T such that

$$\vdash_L A \ \Leftrightarrow \ \exists p \vdash_T p : A.$$

Formal proof-verification for a statement A then consists of the construction of a (fully formalized) proof p (a so-called *proof-object*) and the mechanical verification of $p : A$. As was emphasized by the Bruijn this should be done by a reliable (hence preferably small) program in order to be methodologically sound.

2.3.3 Axiomatization of Type Theory: Pure Type Systems

As was stated before, there is an axiom in type theory that states directly the existence of the generalized Cartesian product: if A is a type and $B(x)$ is type (with parameter $x{:}A$), then $\Pi x{:}A.B(x)$ is a type. If $B(x)$ does not depend on x (i.e. if there is no occurrence of x in $B(x)$), then $\Pi x{:}A.B(x) = A \to B$ ($= B^A$ in set-theoretic notation,[35]) the type of functions from (the inhabitants of) A to (the inhabitants of) B.

Types come in 'kinds'. For example, in some type theories there is a kind $*^s$ whose inhabitants are those types that denote 'sets', and a kind $*^p$ for those that types that denote 'propositions'. The kind $*^s$ is therefore the 'super type' of ordinary types intended to denote sets. But also these kinds appear in various version, so we need 'super kinds', etc. To fix terminology, these are called collectively *sorts*, including the lower kinds like $*^s$. The various type theories differ as to what is allowed as Cartesian product. If s_1, s_2, s_3 are sorts, then

$$\frac{\Gamma \vdash A : s_1 \quad \Gamma, x{:}A \vdash B : s_2}{\Gamma \vdash (\Pi x{:}A.B) : s_3} \quad \text{product } (s_1, s_2, s_3)$$

is called the product-rule, parametrized by the three sorts. This rule works in collaboration with the application and abstraction rules

$$\frac{\Gamma \vdash F : (\Pi x{:}A.B) \quad \Gamma \vdash a : A}{\Gamma \vdash (F\ a) : B[x{:}=a]} \quad \text{application}$$

$$\frac{\Gamma, x{:}A \vdash b : B \quad \Gamma \vdash (\Pi x{:}A.B) : s}{\Gamma \vdash (\lambda x{:}A.b) : (\Pi x{:}A.B)} \quad \text{abstraction}$$

The most simple type theory has just one sort $*$ and as product rule $(*, *, *)$. This is a simplification, due to Ramsey[36] (1903–1930), of Russell's ramified theory of types with sorts $\{*_n \mid n \in \mathbb{N}\}$ and product rules $(*_n, *_m, *_{\max(n,m)})$, see [81].

In type theories with *dependent types*, [47], one allows types to 'depend' on elements of (other) types. An intuitive example is the vector space $\mathbb{R}^n$ with $n \in \mathbb{N}$. This kind of type theory has as sorts $*, \square$ with $* : \square$ and as product rules $(*, *, *)$

---

[35]This is similar to the arithmetic statement $\prod_{i=1}^{3} b_i = b_1.b_2.b_3 = b^3$ if $b_1 = b_2 = b_3 = b$.

[36]As Peter Aczel pointed out to me this simple type theory was already present in Frege.

and $(*, \square, \square)$. In type theories with *higher order types*, see [60], one has the same two sorts, but now as rules $(*, *, *)$ and $(\square, *, *)$ (second order) or $(*, *, *)$, $(\square, *, *)$ and $(\square, \square, \square)$ (higher order). In second order type theory one has for example the inhabited type

$$(\lambda \beta: * \, \lambda x: (\Pi \alpha: * . \alpha).x\beta) : (\Pi \beta: * .(\Pi \alpha: * . \alpha) \to \beta),$$

corresponding to the statement that from the false statement anything follows (*ex falso sequitur quodlibet*). In the calculus of constructions, which underlies the proof assistant Coq, one has as product rules $(*, *, *)$, $(*, \square, \square)$, $(\square, *, *)$ and $(\square, \square, \square)$.

A faithful description of predicate logic is given by the type theory λPRED, with sorts $\{*^s : \square^s, *^p : \square^p\}$ and rules $(*^p, *^p, *^p)$ for implication, $(\square^s, *^p, *^p)$ for quantification, $(*^s, \square^p, \square^p)$ for the formation of predicates.

Definition 1. The *specification of a* PTS *(pure type system)* consists of a triple $S = (\mathcal{S}, \mathcal{A}, \mathcal{R})$ where

1. $\mathcal{S}$ is a subset of the constants C, called the *sorts*;
2. $\mathcal{A}$ is a set of *axioms* of the form $c : s$ with $c \in C$ and $s \in \mathcal{S}$;
3. $\mathcal{R}$ is a set of rules of the form (s_1, s_2, s_3) with $s_1, s_2, s_3 \in \mathcal{S}$. The rule (s_1, s_2) is an abbreviation of (s_1, s_2, s_2).

Definition 2. The PTS determined by the specification $S = (\mathcal{S}, \mathcal{A}, \mathcal{R})$, notation $\lambda S = \lambda(\mathcal{S}, \mathcal{A}, \mathcal{R})$, is defined as follows. Expressions $\mathcal{T}$ are given by the following abstract syntax

$$\mathcal{T} = V \mid C \mid \mathcal{T}\mathcal{T} \mid \lambda V{:}\mathcal{T}.\mathcal{T} \mid \Pi V{:}\mathcal{T}.\mathcal{T}$$

A *statement* (with subject M) is of the form $M : A$, with $M, A \in \mathcal{T}$. A context Γ is an ordered sequence of statements with as subjects distinct variables. $\langle \rangle$ denotes the empty sequence. The notion of type derivation $\Gamma \vdash_{\lambda S} A : B$ (we often just write $\Gamma \vdash A : B$) is defined by the following axioms and rules (Fig. 9). Here $=_R$ is a conversion relation corresponding to a notion of reduction R including at least β, i.e. one always has

$$(\lambda x.M)N \ =_R \ M[x := N]$$

and the deductive power of equational logic (making $=_R$ an equivalence relation compatible with application and λ- and Π-abstraction).
Examples of PTSs.

  (i) $(\lambda \to, [37])$ The simply typed lambda calculus with one ground type o can be specified as PTS as follows.

$$\lambda \to \begin{array}{|c|l|} \hline \mathcal{S} & *, \square \\ \mathcal{A} & o : *, * : \square \\ \mathcal{R} & (*, *) \\ \hline \end{array}$$

(axioms)	$\langle\,\rangle \vdash c : s,$	if $(c:s){\in}\mathcal{A};$
(start)	$\dfrac{\Gamma \vdash A : s}{\Gamma, x : A \vdash x : A}\,,$	if x is fresh;
(weakening)	$\dfrac{\Gamma \vdash A : B \quad \Gamma \vdash C : s}{\Gamma, x : C \vdash A : B}\,,$	if x is fresh;
(product)	$\dfrac{\Gamma \vdash A : s_1 \quad \Gamma, x{:}A \vdash B : s_2}{\Gamma \vdash (\Pi x{:}A.B) : s_3}\,,$	if $(s_1, s_2, s_3){\in}\mathcal{R};$
(application)	$\dfrac{\Gamma \vdash F : (\Pi x{:}A.B) \quad \Gamma \vdash a : A}{\Gamma \vdash Fa : B[x := a]}\,;$	
(abstraction)	$\dfrac{\Gamma, x{:}A \vdash b : B \quad \Gamma \vdash (\Pi x{:}A.B) : s}{\Gamma \vdash (\lambda x{:}A.b) : (\Pi x{:}A.B)}\,;$	
(conversion)	$\dfrac{\Gamma \vdash M : A \quad \Gamma \vdash B : s}{\Gamma \vdash M : B}\,,$	if $\quad A =_R B$

Fig. 9 The PTS $\lambda_R(\mathcal{S}, \mathcal{A}, \mathcal{R})$

(ii) ($\lambda 2$, also called *system F*, [60]) The second order polymorphic lambda calculus can be specified as PTS as follows.

$$\lambda 2 \quad \begin{array}{l|l} \mathcal{S} & *, \square \\ \mathcal{A} & * : \square \\ \mathcal{R} & (*, *), (\square, *) \end{array}$$

(iii) (The λ-cube, [11]) The calculus of constructions as a PTS is specified by

$$\lambda C \quad \begin{array}{l|l} \mathcal{S} & *, \square \\ \mathcal{A} & * : \square \\ \mathcal{R} & (*, *), (*, \square), (\square, *), (\square, \square) \end{array}$$

(iv) λPRED, logic as a PTS, is determined by the following specification.

$$\lambda\text{PRED} \quad \begin{array}{l|l} \mathcal{S} & *^s, *^p, *^f, \square^s, \square^p \\ \mathcal{A} & *^s : \square^s, *^p : \square^p \\ \mathcal{R} & (*^p, *^p), (*^s, *^p), (*^s, \square^p), \\ & (*^s, *^s, *^f), (*^s, *^f, *^f) \end{array}$$

(v) (The inconsistent type theory $* : *$)

$$\lambda * \begin{array}{|cc|} \hline \mathcal{S} & * \\ \mathcal{A} & * : * \\ \mathcal{R} & (*, *) \\ \hline \end{array}$$

(vi) The Curry-Howard 'isomorphism' is the map $\theta : \lambda\text{PRED} \rightarrow \lambda C$ given by

$$\theta(*^i) = *$$
$$\theta(\square^i) = \square$$

See [11] for a discussion of such and similar 'Pure Type Systems'. These were introduced by Berardi and Terlouw as a generalization of the lambda cube.

2.3.4 Inductive Types

For representation of mathematics we need also inductive types. These are freely generated data types like

```
nat   :=   O:nat  |  S:nat→nat
```

for the representation of freely generated data types. Inductive types come together with terms for primitive recursion and at the same time induction.

```
nat_rec_s :   ΠP:(nat->s).
                ((P O)->(Πn:nat.(P n)->(P (S n)))->(Πn:nat.P n))
```

for which the following rewrite rules are postulated.

```
nat_rec_s P a b O       →_ι  a
nat_rec_s P a b (S n)   →_ι  b n (nat_rec_s P a b n)
```

Also predicates can be defined inductively.

```
le [n,m:nat]   :=   le_n : (Πn:nat.(le n n) |
                    le_S : (Πn,m:nat.((le n m)→(le n (S m))))).
```

This means that we have the axiom and rule (writing $n \leq m$ for $(\text{Lt } n \ m)$)

$$\frac{}{n \leq n} \ (\text{le\_n } n) \qquad \frac{n \leq m}{n \leq (S \ m)} \ (\text{le\_S } n \ m) .$$

Properties of $\leq$, like transitivity, can be proved by induction on the generation of this inductive relation.

Inductive types in the context of type theory have been first proposed in [108]. Their presence makes formalization easier. A technical part of Gödel's

incompleteness proof is devoted to the coding of finite sequences of natural numbers via the Chinese remainder theorem. Using an inductive type for lists makes this unnecessary. See [50, 98] and [32] for a formal description of inductive types, related to those used at present in the mathematical assistant Coq.

2.3.5 The Poincaré Principle

The reduction rules generated by the primitive recursion ι-contractions, together with β-reduction from lambda calculus and δ-reduction (unfolding definitions) play a special role in type theory.

$$\frac{\Gamma \vdash p : A \quad \Gamma \vdash B : s}{\Gamma \vdash p : B} \quad A =_{\beta\delta\iota} B$$

This has, for example, the consequence that if p is a proof of $A(4!)$, then that same p is also a proof of $A(24)$. This is called the *Poincaré Principle*[37] for $\beta\delta\iota$-reduction, see [12, 49]. It has a shortening effect on the lengths of proofs, while in practice the simple decidability of $p : A$ is not much increased. In fact

$$\Gamma \vdash p : A \iff \text{type}_{\Gamma}(p) =_{\beta\delta\iota} A.$$

Now in principle already the complexity of $=_{\beta}$ is expensive. In fact it is PSPACE complete, as [111] has shown. But for terms coming from humans it turns out that this complexity upper limit is far from being used in natural situations (i.e. in from proofs given by hand).

The Poincaré Principle can be postulated for different classes $\mathcal{R}$ of rewrite relations. Below we will discuss that in some mathematical assistants (HOL) $\mathcal{R} = \emptyset$, in others (PVS) essentially one has that $\mathcal{R}$ corresponds to a large set of decision procedures.

2.3.6 Propositions-as-Types

The already mentioned propositions-as-types interpretation was first hinted at by in [45], and later described by [70]. It is related to the intuitionistic interpretation of propositions. A proposition A is interpreted as the collection (type) of its proofs

---

[37] Speaking about "$2 + 2 = 4$" and its proof in some logical system, [101], p. 12, states: *Mais interrogez un mathématicien quelconque: "Ce n'est pas une démonstration proprement dite", vous répondra-t-il, "c'est une vérification."* .

$[A] = \{p \mid p$ is a proof of $A\}$. Martin-Löf [88] completed the interpretation by showing how inductive types can be used to give a very natural interpretation to this.

```
A× B       :=   pair:A→B→(A× B)
A+B        :=   in left:A→(A+B)|in right:B→(A+B)
Ø          :=
Σx:X.B     :=   pair:Πx:X.(B→Σx:X.B)
```

Using these one has the following.

$$[A \rightarrow B] := [A] \rightarrow [B]$$

$$[A \ \& \ B] := [A] \times [B]$$

$$[A \vee B] := [A] + [B]$$

$$[\bot] := \emptyset$$

$$[\forall x{:}X.B] := \Pi a{:}X.[B[x:=a]]$$

$$[\exists x{:}X.B] := \Sigma a{:}X.[B[x:=a]]$$

Predicative type theory has been argued by Martin-Löf to be the right kind of foundation.

2.3.7 Category Theory

Category theory can be axiomatized in a two sorted predicate logic (one for objects and another for arrows) with partial terms (composition of arrows is not always defined) and equational logic. For every object A there is an arrow $\mathrm{id}(A)$ with $\mathrm{dom}(\mathrm{id}(A)) = \mathrm{cod}(\mathrm{id}(A)) = A$; for every arrow f there are objects $\mathrm{dom}(f)$ and $\mathrm{cod}(f)$. If f, g are arrows and $\mathrm{cod}(f) = \mathrm{dom}(g)$, then $g \circ f$ (sometimes also written as $f; g$) is defined and $\mathrm{dom}(g \circ f) = \mathrm{dom}(f)$, $\mathrm{cod}(g \circ f) = \mathrm{cod}(g)$. The following axioms hold $f \circ (g \circ h) = (f \circ g) \circ h$, $\mathrm{id}(\mathrm{dom}(f)) \circ f = f$ and $f \circ \mathrm{id}(\mathrm{cod}(f)) = f$. In a diagram one has $(f, g, h, \ldots$ range over the arrows, $A, B, \ldots$ over the objects and $g \circ f \downarrow$ means that $g \circ f$ is defined) the following. If an equation is postulated it has to be interpreted as: if the LHS is defined then so is the RHS and conversely and in either case both are equal (Fig. 10).

The equational reasoning is often done in a pictorial way (using diagrams). For an implication like

$$g \circ f = k \circ h, \quad q \circ p = r \circ k \ \Rightarrow \ r \circ g \circ f = q \circ p \circ h$$

Fig. 10 The axioms of
category theory

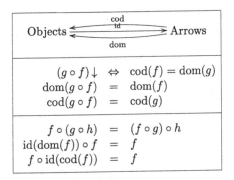

$$(g \circ f)\downarrow \quad \Leftrightarrow \quad \mathrm{cod}(f) = \mathrm{dom}(g)$$
$$\mathrm{dom}(g \circ f) \quad = \quad \mathrm{dom}(f)$$
$$\mathrm{cod}(g \circ f) \quad = \quad \mathrm{cod}(g)$$

$$f \circ (g \circ h) \quad = \quad (f \circ g) \circ h$$
$$\mathrm{id}(\mathrm{dom}(f)) \circ f \quad = \quad f$$
$$f \circ \mathrm{id}(\mathrm{cod}(f)) \quad = \quad f$$

one draws a commutative diagram:

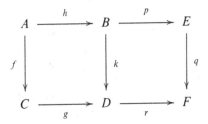

2.3.8 Category Theory as Foundation

One can view category theory as a foundation for mathematics. By imposing extra axioms, e.g. there exists only one object, the category under consideration becomes a particular mathematical structure (in this case a monoid). This method is strong enough to be a foundation for close-up mathematics. By imposing that we work in a more complex category, e.g. a topos that has a rich structure of objects and arrows, one obtains a wide-angle mathematical view with many (internal) categories in which the various mathematical structures that exists can be embedded and compared. In this respect category theory makes little ontological commitment: things are valid depending on the particular category one starts with.

The interpretation of an intuitive mathematical statement "Given situation Γ, then one has A" can be interpreted both in the way as in logic or as in type theory. In the first case the (translation of the) statement A becomes valid in a category enjoying some properties depending on Γ; in the second case one takes into account the proof p of A and validity becomes (after translation into the right *fibered* category) $p(A) = 1$, see [73].

For some of the categories, e.g. the effective topos and several of the fibered categories, the elements have a computational flavor. This, however, has not been exploited in a system for computer mathematics. On the other hand the functional

programming language Ocaml,[38] is implemented on the Categorical Abstract Machine, [42], inspired by category theory. Moreover, Ocaml is the implementation language of Coq.

2.3.9 A Comparison Between the Three Foundations

Categories plays a mediating role in the foundations of mathematics. Some of the type theories are quite wild, notably the higher order ones. By providing category theoretic semantics for type theory, see [73], some clarity can be obtained.

As the needed categories do exist within set theory, one has bridged the "ontological gap" between set theory and type theory: if one believes in set theory, one also ought to believe in type theory. Aczel argues that one should do the opposite, i.e. found the reliability of set theory on type theory. In a series of papers [2–4] and [5] CZF (constructive (predicative) ZF set theory) is based on the predicative type theory of [88]; also (the much stronger) IZF on a type theory containing an analogue of the sub-object classifier in a topos.[39]

As a foundation for mathematics category theory stands between set theory and type theory. It needs equational logic in order to deal with equality between expressions consisting of composed arrows (using composition). In this respect it is much like most type theories that also need equational logic. But in a category one sometimes wants to state e.g. the existence of a pull-back and for these one needs quantifiers. Type theory, on the other hand, does not need logic using quantifiers; these are built in.

The analogy with set theory is somewhat different. Both category and set theory use first order logic with equality. Both have enough ontological expressive force to be able to describe the necessary spaces in mathematics. There is a noticeable difference in that category theory (like group theory) is polyvalent in intension: there are many categories (and groups). Set theory has a fixed ontological commitment: it is (or at least, was originally) intended to describe the structure of the platonic universe. But it does not succeed in this, as there are many independent statements like $2^{\aleph_0} = \aleph_1$, even under the assumption of extra large cardinal axioms. Given this state of affairs, for the foundation of mathematical theorems it may be equally arbitrary to choose a model of set theory as choosing a topos. Relativity is also evidenced in type theory by its many possible choices of sorts (universes), product rules and sets $\mathcal{R}$ for which the Poincaré Principle holds. On the other hand in the predicative Martin-Löf type theory a very precise ontological commitment is made.

In set theory mathematical objects have a specific 'implementation', namely as a set. This may cause unnatural questions for example whether the real number 0 is

---

[38] ⟨caml.inria.fr/ocaml⟩

[39] It is open whether IZF can be interpreted in some extension of the calculus of constructions with inductive types and universes.

an element of the real number π.[40] In the category and type theoretic foundations one has a structuralist point of view: one does not say what 0 or π are, only what relations they form with other mathematical objects. This is a strong point in favor of category and type theory. Therefore category theory (and one could add type theory) is sometimes said to conform to the structuralist view of mathematics; see [86]. Also in set theory with Ur-elements like KPU, see Barwise[41] [20], there is a more structuralist point of view.

3 Foundational Views and Criticism

3.1 Formalism

In Sect. 1 it was mentioned that mathematics consists of a tower of theories with at the bottom elementary ones like geometry and arithmetic. In the higher theories notions of an infinite nature play a definite role. One considers sets like $\mathbb{N}, \mathbb{N} \rightarrow \mathbb{N}$ and $\mathcal{P}(\mathbb{N})$ as a completed entities, since one may quantify over them. For example

$$D = \{n \in \mathbb{N} \mid \exists x, y \in \mathbb{N}. \, 7y^2 = nx^3 + 1\}$$

The existence of such sets D seems to imply that we know them.[42] But in fact such Diophantine[43] sets are in general undecidable. Even more problematic is the following notion.

$$P(n) \quad \Leftrightarrow \quad \forall X \subseteq \mathbb{N}. \, [n \in X \ \& \ \dots \ \Rightarrow \ 0 \in X].$$

This is called an impredicative definition, because P, essentially a subset of $\mathbb{N}$ is defined in terms of all possible subsets of $\mathbb{N}$. So in this case the property $P(n)$ depends on all possible subsets of $\mathbb{N}$, including the one determined by P.

The question arose whether the results that can be proved from these higher ("*infinitary*") concepts remain provable without them ("*by finitistic means*"). In order to achieve this Hilbert set out a program to establish this. In [109] this is described as follows. Hilbert was not precise as to what is a theory with infinitary character and what is a finitary theory. Simpson argues convincingly that second order arithmetic Z_2 and primitive recursive arithmetic PRA may be seen as a good

---

[40]In the system for Computer Mathematics Mizar, see Sect. 5, the answer is "Yes", but the answer depends on the fact that π is irrational.

[41]1942–2000.

[42]For $n \in D$ one has at least as possibilities $6, 27, 62, 111, 174, \dots, 7k^2 - 1, \dots$ (take $x = 1$), and also $1, 162, 218, 701$ and 813 (take $x = 3$), but it is not immediate whether there are infinitely many more solutions.

[43]After Diophantos of Alexandria (app. 200–284 AD)

operationalizations of these concepts respectively. Indeed most of mathematics may be formalized in Z_2, while PRA is definitely an innocent theory, as it has no explicit quantifiers. In this view Hilbert's program wants to establish that statements in PRA provable in Z_2 are already provable in PRA. This is called conservativity (of Z_2 over PRA). The way Hilbert wanted to achieve this is nowadays called reflection. By seeing provability as a game with symbols one could describe it within PRA and then establish mentioned conservativity.[44]

As Hilbert's program was the first serious attempt of reflection via a formal description of mathematics, he was said to adhere to the formalist philosophy: mathematics devoid of any meaning. I think this is unfair, as Hilbert was very much aware of the meaning of finitistic mathematics and just wanted to establish the same for "higher" mathematics.

3.2 Logicism

Whereas in formalism mathematics is reduced to a game with symbols, in logicism the rules of logic are seen as having a meaning and being valid. According to this view mathematics developed using the axiomatic method thereby follows logic. This seems plausible, as (first order) logic proves all statements that hold in all models of the axioms (completeness, see [61, 110]). Traditionally Frege and Russell are considered as logicists. When Frege had completed the quest for logic initiated by Aristotle, he started to formalize mathematics within his system. For this he had to use some ontology and his choice here was Cantorian set theory that he gave a particular logical basis. The axioms of set existence, mentioned in Sect. 2, all could be postulated by the comprehension axiom:

$$\exists x \forall y.[y \in x \leftrightarrow P(y)],$$

for arbitrary formulas $P(y)$. Frege's system was shown to be inconsistent by Russell, who together with Whitehead (1861–1947) started a different formal system, some form of Type Theory, in which (parts of) mathematics could be formalized. Although the logical aspects of this system where somewhat sloppy (they failed to treat the difference between free and bound variables, forcing the reader to make systematic disambiguations), they succeeded for the first time in history to provide a formalized version of number theory. So actually Russell and Whitehead (and also Frege) could have been called formalists as well. The impact of Principia Mathematica was not mathematical, but metamathematical. Since the

---

[44]Gödel's famous incompleteness result showed that conservativity does not hold. But Simpson argues that Hilbert's program is partly successful. He estimates that about 85% of mathematics can be proved in the system PRA+WKL$_0^+$, where WKL$_0^+$ is some form of König's Lemma stating that a finitely branching infinite tree has an infinite path, and this system *is* conservative over PRA.

collection of provable statements in arithmetic in the context of Principia became a well-defined notion, one could wonder whether it was complete (for each numerical sentence A one has that either A or $\neg A$ is provable) and decidable (a machine could decide whether A is provable). Both questions turned out to have negative answers as we learned from Gödel on the one hand and Church and Turing on the other hand. The Gödel incompleteness results in this respect should be seen as a limitation of the axiomatic method (but we do not have something better).

Serious arguments against logicism came from Poincaré and Skolem. In order to prove that $2 + 2 = 4$ one needs very many steps in a logicistic system like Peano arithmetic formulated in first order predicate logic. As mentioned before, Poincaré stated that this is "just a verification." The Poincaré Principle in type theory aims to overcome this.

3.3 Platonism

In Platonism the mathematical objects are taken for real. Logic is only a way to reveal some of the truths about these real objects. Set Theory is a theory that describes these objects. Mathematics is a walk through the paradise of Cantor. One sees better and better what theorems are valid in this world. One consequence of the belief in the existence of actual infinity is the belief in the following principle[45]:

$$\neg\neg\exists x{:}\mathbb{N}.\,P(x) \;\Rightarrow\; \exists x{:}\mathbb{N}.\,P(x).$$

Indeed, if $\mathbb{N}$ exists as a totality, and if P is a well-defined property over $\mathbb{N}$ that can be represented as a subset, then one just needs to see whether there is an element of this subset. If there is none, then we obtain a contradiction. So there must be an element! Of course, this reasoning is circular, as it depends on the double negation law (closely related to the excluded third), that we basically want to prove. But the reasoning shows how compelling it is.

Criticism against the Platonist view has been formulated by Feferman in a series of papers collected in [52]

  (i) Abstract entities are assumed to exist independently of any means of human definition or construction;
 (ii) Classical reasoning (leading to nonconstructive existence results) is admitted, since the statements of set theory are supposed to be about such an independently existing reality and thus have a determinate truth value (true or false);
(iii) Completed infinite totalities and, in particular, the totality of all subsets of any infinite set are assumed to exist;

---

[45]For decidable P this is called *Markov's Principle*, after A.A. Markov Jr. (1903–1979).

(iv) In consequence of (iii) and the Axiom of Separation, impredicative definitions of sets are routinely admitted;

(v) The Axiom of Choice is assumed in order to carry through the Cantorian theory of transfinite cardinals.

Mostowski (1968, personal communication) once said: "Peano arithmetic is probably consistent. Impredicative set theory may very well be inconsistent. In that case the inconsistency may already be present in second order arithmetic. Also the notions of forcing play already a role in this theory. That is why I like to work in this field.[46]" One should add, that in this respect Girard's system $\lambda 2$, see [60] where it is called 'System F', is very interesting. Its strong normalization is equivalent to the consistency of second order arithmetic and can be proved using impredicative methods. But since these principles are dubious, the strong normalization is not so reliable as that of the system T of Gödel, see [114]. In fact, for the system $\lambda*$ (the PTS with $* : *$ and $(*, *)$) strong normalization was proved using the methods of the system itself, while the system is not strongly normalizing, see [72] simplifying a proof of Girard.

One of the advantages of (impredicative) set theory is that it is so strong that it is able to embed all results from most other foundations. This gives mathematics a unity, as has been emphasized by Girard. (The exception consists of those parts of Intuitionism depending on continuity principles and other non-classical statements to be discussed below. But there are classical interpretations of those parts of intuitionism.)

3.4 Calculism

In Calculism the emphasis is put on computing. At first there was the overly optimistic belief in the power of computing by Leibniz as discussed above. Even at the dawn of the twentieth century Hilbert believed that e.g. solvability of Diophantine equations could be decided in a computable way (In [68] his 10-th problem was to establish an algorithm to do just that; but as mentioned above this is impossible.) After the notion of general computability had been captured, it was proved by [36] and [116] that validity or provability of many mathematical questions was not decidable. Certain important theories are decidable, though. Tarski [113] showed that the theory of real closed fields (and hence elementary geometry) is decidable. An essential improvement was given by [39]. In [29] a method to decide membership of finitely generated ideals in certain polynomial rings was developed. For polynomials over $\mathbb{R}$ this can be done also by the Tarski-Collins method, but much less efficiently so. Moreover, "Buchberger's algorithm" was optimized by

---

[46]Later he added: "But when the secret police comes to ask on what grounds I choose my object of research, then I tell them: 'I study it, because the Americans study it'; and the next year I say: 'I study it, because the Russians study it'!"

several people, e.g. [9]. It has impressively many applications ranging from the fields of robotics to differential equations, see [30, 106].[47] Another use of calculism is the automated theorem prover for geometry, see [34], based on translating putative theorems into algebra, where they are decided by manipulating polynomial inequalities. See also [27] for other examples of decidable theories. It is the success of calculism, even if partial, that has been neglected in mathematics based on set theory or logic alone.

3.5   Intuitionism

Intuitionism is a foundational view initiated by Brouwer[48] [28]; see [46] and [115] for later developments. It proposes to use a sharper language than in Classical Mathematics. Brouwer's observation was the following. If one claims that $A \vee B$ holds one also wants to be able indicate which one is the case. Similarly, if $\exists x.A$ holds, then one wants to be able to find a 'witness' and a proof of $A[x := a]$. In classical mathematics, based on Aristotelian logic, this is not always the case. Take for example the arithmetic statement (where GC is an open problem like the Goldbach Conjecture)

$$P(x) \;\Leftrightarrow\; (x = \underline{0} \;\&\; \text{GC}) \vee (x = \underline{1} \;\&\; \neg\text{GC}).$$

Now one can prove $\vdash \exists x.P(x)$ (indeed if GC holds take $x = 0$ else $x = 1$), without having $\vdash P(\underline{0})$ or $\vdash P(\underline{1})$ because the GC is still open (and certainly one has $\nvdash P(\underline{n})$ for $n > 1$). Similarly one has $\vdash$ GC $\vee \neg$GC, without having $\vdash$ GC or $\vdash \neg$GC. One may object that sooner or later GC may be settled. But then one can take instead of GC an independent Gödel sentence that for sure will not be settled (if arithmetic is consistent). Brouwer analyzed that this imperfection was caused by the law of excluded middle $A \vee \neg A$. Heyting formulated a logical system (intuitionistic[49] predicate logic) that remedied this effect, as proved by Gentzen (who also gave intuitionistic logic a nicer form: see the system in Fig. 2 leaving out the double negation rule from classical logic).

   Another criticism of Brouwer (against the logicistic view this time) is that logic does not precede mathematics. For example if one wants to formulate logic, one needs to define the context free language of formulas. This criticism has been dealt

---

[47]The Buchberger algorithm and that of [79] are closely related, see [91] and [87].

[48]1881–1966.

[49]It is interesting that there is a set theoretic semantics of intuitionistic propositional logic comparable to that of the classical version. The latter theory can be interpreted in Boolean algebras with as prototype subsets of a given set. The intuitionistic theory as Heyting algebras with as prototype the open subsets of a topological space X. Negation is interpreted as taking the interior of the complement, disjunction as union. And indeed in general one does for $A \subseteq X$ that $A \cup \overline{A}^{\,o} \neq X$. Therefore the law of the excluded middle fails in this model.

with somewhat in type theory where next to the logical axioms there are axioms concerning data types.

The intuitionistic school of mathematics at first did not gain much interest. One of the reasons was that mathematics had to be reproved and possibly modified. Now that this work has obtained a serious start one collects the fruits. In set theory a theorem like

$$\forall n \in \mathbb{N} \exists m \in \mathbb{N}.P(n,m) \tag{3}$$

does not imply that the m can be found computably form n. If one wants to express this computability, then it is not even enough to state and prove

$$\exists f \text{ computable } \forall n \in \mathbb{N}.P(n, f(n)),$$

as the $\exists f$ may not lead to a witness for a computable function. The only way to state that in (3) the m is computable in n is to actually give the algorithm, which in set theory is not very practical. In intuitionistic mathematics provability of (3) automatically implies computability (and if computability does not hold one can reformulate (3) as

$$\forall n \in \mathbb{N} \neg\neg\exists m \in \mathbb{N}.P(n,m). \tag{4}$$

For these reasons (R. L. Constable, 1997, personal communication) stated "Intuitionism nowadays has become technology". A challenging subject is to extract programs from fully formalized $\forall\exists$ statements, see [85, 97]. Although this is possible in principle, there is space for optimizations. As pointed out by [80], see also [107] and [43], the information of proofs of negative statements is irrelevant, so that these need to be discarded. Moreover, in several cases classical proofs can be transformed into intuitionistic proofs (for example if the statement is an arithmetic Π_2^0 statement, see [41, 55]) and widens the scope of the extraction technology, see [2].

3.5.1 Constructivism vs Intuitionism

Brouwer not only criticized the double negation law, he also stated principles that contradict it, for example that all functions $\mathbb{R} \to \mathbb{R}$ are continuous,[50] see [115], Chap. 4. Constructivism consist of the part of intuitionism by just leaving out the double negation law, see [26, 93]. Although it seems daring to state axioms contradicting classical mathematics, one should realize that with some effort much

---

[50]A classical function that contradicts this is the step function $s(x)$ that is 0 for $x < 0$ and 1 otherwise. But intuitionistically s is not definable as total function, as one cannot determine from e.g. a Cauchy sequence whether its limit is < 0 or ≥ 0.

of classical mathematics can be reformulated in a way such that it becomes constructively valid. This means that there is place for extensions like the continuity theorem. The strong intutionistic principles then can be seen as a welcome additional power. For those functions that are provably total, one can show that they are continuous indeed. This was how Bishop (1928–1983) understood Brouwer's explanation of these axioms, see [25]. Finally it should be observed that in mathematical models occurring in physics, e.g. microelectronics, all total discontinuous functions like 'square waves' are just a *façon de parler*.

4 Computer Mathematics

Computer Mathematics (CM) is mathematics done by humans and computers in collaboration. The human gives the development of a theory: definitions and theorems. The computer checks whether the definitions and statements of theorems are well formed; then in an interactive fashion some or all steps of the proof are given; finally the computer checks their correctness. One purpose of CM is to assist teaching existing mathematics and develop new mathematics. Another purpose is to reach the highest degree of reliability. Last but not least, through CM it will be possible to have a certified library of theories, ready for reuse and exchange, see [15,38]. At present CM is not yet established, but forms an interesting challenge. See [22] for a stimulating discussion of the subject, with examples not covered in this paper.

One part of CM is Computer Algebra (CA). It deals with 'computable' objects, often in an equational way. This by now is an established, though developing, subject. Mathematical Assistants deal with side conditions of equations and more general with reasoning that cannot be formulated in CA.

4.1 Computer Algebra

Systems of CA, like the commercial packages Maple and Mathematica or the systems more directed toward mathematical research like Gap, Li, Magma and Pari, all represent "computational" mathematical objects and helps the human user to manipulate these. Computational objects not only consists of numbers and polynomials, but also of expressions built up from roots, integrals, transcendental functions, for example the elliptic integral of the first kind

$$f(\alpha) = \int_0^\alpha \frac{1}{\sqrt{1 - \frac{1}{4}\sin^2 \varphi}} \, d\varphi.$$

The more advanced systems (like Magma) represent groups, e.g. $\text{Aut}_{\mathbb{F}_7}(C)$, with C being the hyper elliptic curve $\{\langle x, y \rangle \in \mathbb{F}_7^2 \mid y^2 = x^4 + 4\}$ over the finite field with

seven elements.[51] That it is possible to represent on a computer an object like $\sqrt{2}$, that has infinitely many digits in its decimal representation, follows from the fact that it can be represented by a single symbol, but we know how to manipulate these symbols. For this reason we call mentioned objects computable.

4.2 Mathematical Assistants

In systems of Computer Mathematics one even can represent arbitrary mathematical notions. Moreover, the systems that can handle these, the mathematical assistants, help the human user to define new mathematical notions and make deductions with them. The reason that the constraint of computability now can be dropped is the following. Even if for a property P and object c it may be undecidable whether $P(c)$ holds, it is decidable whether a putative proof p of this statement is a valid proof indeed.

4.3 Formal Systems

There is a choice of formal system in which mathematics is being represented. Frege made a start, but when formalizing Cantorian set theory in (his) predicate logic, the system unfortunately became inconsistent as shown by Russell through his paradox. Then Russell and Whitehead chose a form of type theory and made a reasonable start. Their description of the theory lacks rigor though.[52] Curry [44] worked with extensions of the untyped lambda calculus, but suffered from either weakness or inconsistencies of the systems involved.[53] Church [37] introduced the theory of simple types that is the basis of the mathematical assistant HOL.

McCarthy [90] made a plea for formalization together with a computer verification, using first order logic. He did not get far, because of the lack of force of this logic (one can represent the close-up theory of statements valid in all groups, but not the wide-angle theory of groups, unless one formalizes set theory) and because of his proposal to represent proofs in the Hilbert way (a sequence of formulas that either are axioms or follow from previous formulas) was cumbersome. But McCarthy had good ideas about combining formal proofs and symbolic computations in the Babylonian style.

---

[51]It writes this group as a product of simpler groups and tells us how the elements act on the points of the curve C.

[52]For example free and bound variables are used in such a way that the reader has to insert in many places a binder. See [81].

[53]Only in the 1990s adequate systems of *Illative Combinatory System* have been given, see [16,48]. In some sense these are simpler than the PTSs, in another sense they are more complicated.

Important progress came from de Bruijn [47], see for a survey [95], with his family of Automath languages and corresponding proof-checkers. These languages all are based on some form of type theory extended by the dependent types already discussed. The admittedly already quite formal "Grundlagen der Analysis" by Landau[54] [82] has been formalized in AUT-68, a member of the Automath family, by [118] and exactly one error was found in it. It was emphasized by de Bruijn emphasized that a mechanical proof-checker should have a small program (so that it can be seen "by hand" to be correct), otherwise the correctness of the verification becomes a point. This is called the *de Bruijn criterion*.

4.4 Poincaré Principle

Another 'parameter' relevant for Mathematical Assistants is the way in which calculations are supported. If the formal system has as derivation rule

$$\frac{A(f(t))}{A(s)} \; f(t) = s,$$

then we say that it satisfies the Poincaré Principle (PP) for f. The class $\mathcal{P}$ of functions for which the Poincaré Principle holds varies over the formal systems for CM. If $\mathcal{P} = \emptyset$, then formal proofs involving computations become quite long. We discussed that Poincaré and Skolem criticized logicism for this reason. In that case the proof-objects become so large (they essentially contain the traces of necessary computations) that they will not be stored. The way these will be checked is still quite reliable. The proofs are being represented bit by bit in a working memory and local correctness is checked. As soon as a part turns out to be correct it is being erased. We speak about *ephemeral proof-objects*. For these systems only the proof-script that generates the ephemeral proof-object will be stored. In [103] a stronger proposal is made by viewing decision methods as admissible rules with (semi) decidable side conditions.

4.5 Reflection

The method of reflection, that had its applications in projective geometry, meta-mathematics, set theory, model theory and category theory, also becomes important in computer mathematics in order to provide formal proofs for statements otherwise obtained by intuition or computation. A particularly fruitful use of reflection is as follows. If we want to prove a property $A(t)$, where A is some predicate and t is

---

[54] 1877–1938.

some term, then sometimes the method of *generalization* simplifies matters: one first proves $\forall x.A(x)$ in order to conclude $A(t)$. The method of *pattern generalization* is more useful. If we want to prove $A(t)$, then we can often write $t = f(s)$ and prove also $\forall x.A(f(x))$. As soon as we can prove $t = f(s)$ (employing the Poincaré Principle or using ephemeral proofs) we are done. The terms s are often of a syntactical nature and the map f involves semantic interpretation $s \mapsto [\![s]\!]$. An example of this use of reflection is the following. In order to prove that the elliptic integral $f(\alpha)$ defined above is continuous in α an attentive student can see this immediately from the defining expression. If the expressions become more complex it is a burden to provide formal proofs of these facts. It can, however, be done in a light way, closely following our intuition. This is done by introducing a formal language L containing expressions for functions like f, together with an interpretation function $[\![\;]\!]$ transforming this expression in the intended actual function. One only needs to prove once and for all

$$\forall e{:}L.[\![e]\!] \text{ is continuous}$$

and then one can apply this to the *quote* of f. For this a provable computation is needed to show $f = [\![\text{quote } f]\!]$, but that can be done either via the Poincaré Principle or ephemeral proofs. In a similar way a computational statement like

$$(xy - x^2 + y^2)(x^3 - y^3 + z^3) = x^4y - xy^4 + xyz^3 - x^5 + \\ x^2y^3 - x^2z^3 + y^2x^3 - y^5 + y^2z^3.$$

can be proved by reflection and primitive recursive computation. See [13] for more details. The first place where reflection occurred in proof-assistants is in [71]. In [18, 19], the kernel of Coq has been reflected in Coq itself.

4.6 Systems

Various systems evolved. In most of them the human user constructs the formal proof (the so called *proof-object*) assisted by the computer, by interactively writing a *proof-script*. The resulting proof-object will be verified by the small proof checker. The reason for the requirement that the system be small, the so-called *de Bruijn criterion*, is that even if a formal proof (a so called *proof-object*) is huge, it is reliable, because we are able to check the correctness of the software that performs the verification by hand. Even so, several of the kernels of the present systems of Computer Verification have had bugs. These were caused by logical inconsistencies, faulty module systems, or more technical defects (Fig. 11).[55] Some of the systems

---

[55]The mechanism of α-conversion (changing names of bound variables) is often implemented wrongly. In [56] this mechanism is described in a very succinct way so that it will be helpful if implementers use this as foundation of their implementation.

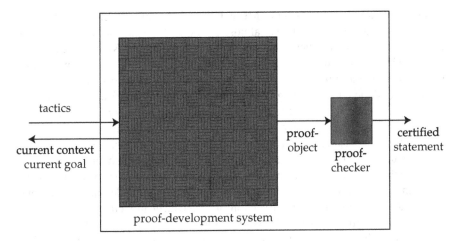

Fig. 11 A mathematical assistant

come with a fixed foundational formal system. Other ones are *logical frameworks*, in which the foundational system may be specified and used.

The main systems for CM in which some substantial theories (i.e. on the order of >10 MB, but still small from a working mathematician's point of view) have been represented are the following.

1. HOL[56] and Isabelle[57];
2. Coq[58] and Nuprl[59];
3. Mizar[60];
4. PVS.[61]

Many other systems can be found at Freek Wiedijk's homepage.[62]

HOL is based on higher-order logic. Its class of functions satisfying the Poincaré Principle is empty. Therefore the proof-objects are huge and therefore made ephemeral and the proof-scripts are being stored. Isabelle is a logical framework that has HOL as one of its main implementations. For this reason the two systems are listed together.

Coq is based on higher order intuitionistic intensional type theory with the Poincaré Principle for $\beta\delta\iota$-reduction. Nuprl is similar, but based on extensional type theory, as discussed in Sect. 2. This has the advantage that subtypes and quotient types can be represented in a natural way and the disadvantage that belonging to

---

[56] ⟨www.cl.cam.ac.uk/Research/HVG/HOL⟩

[57] ⟨www.cl.cam.ac.uk/Research/HVG/Isabelle⟩

[58] ⟨pauillac.inria.fr/coq⟩

[59] ⟨www.cs.cornell.edu/Info/Projects/NuPrl/nuprl.html⟩

[60] ⟨www.mizar.org⟩

[61] ⟨pvs.csl.sri.com⟩

[62] ⟨www.cs.kun.nl/~freek/digimath/index.html⟩

a type requires a proof obligation. Mizar is based on ZF with choice and some large cardinal assumptions. It has almost no Poincaré Principle and cannot do interesting computations (there is no support for ephemeral proofs). Also the attempt of Bourbaki to found mathematics on set theory suffered from this and eventually it did lead to the termination of that project. PVS is based on primitive recursive arithmetic and the corresponding Poincaré Principle and is userfriendly for not too abstract mathematics. One nice feature of Mizar is that it has a *mathematical mode*: proofs are relatively close to the informal proofs. This feature has been added to a variant of Isabelle, Isar, and is presently in consideration for Coq, see [14].

Case studies in Coq include a constructive development of mathematical analysis including the Fundamental Theorems of Algebra and Analysis, the correctness of an algorithm for obtaining Gröbner bases and the FFT and the key lemma for the Four Color Theorem using a compiled version of the reducer in Coq (which has been verified in the simpler interpreted version). Moreover primality of 40 digit numbers has been established by Pocklington's criterion and using the Computer Algebra System Gap for the factorization and congruences (Gap gave input that has been fully verified in Coq).

5   Foundations from a Computer Mathematics Perspective

There is a view of Randy Pollack (1994, personal communication) on formalism, logicism and intuitionism, that gives them a different perspective. These three philosophies correspond to the way mathematics is represented in a Mathematical Assistant and depend on which side(s) of the triangle obtains received ample attention in the basic checker while mathematical activities the other sides have to be provided by the person who formalizes mathematical texts.

5.1   Formalism

If the basic program is such that all aspects of mathematics (defining, reasoning and computing) have to be programmed into the underlying proof assistant, then one may speak of a mathematical assistant in the formalist style. Examples of such assistants are the original Automath system and Isabelle[63] in which simple things like implication have to be programmed by the user. The following is a piece of text in AUT-68 (Fig. 12).

---

[63]This system is a so-called logical framework, i.e. designed in order to represent arbitrary concrete formal systems, even ones that are meaningless.

```
1.  BOOLEANS

1.1            @   bool       :=    PN                             :    'type'
1.2            @   x          :=    ---                            :    bool
1.3     x      @   TRUE       :=    PN                             :    'type'
1.4            @   CONTR      :=    [v:bool]TRUE(v)                :    'type'
1.5            @   a          :=    ---                            :    CONTR
1.6     a      @   b          :=    ---                            :    bool
1.7     b      @   then_1     :=    <b>a                           :    TRUE(b)
1.8            @   ksi        :=    ---                            :    'type'
1.9     ksi    @   nonempty   :=    PN                             :    bool
1.10    ksi    @   a          :=    ---                            :    ksi
1.11    a      @   then_2     :=    PN                             :    TRUE(nonempty)
1.12    ksi    @   a          :=    ---                            :    TRUE(nonempty)
1.13    a      @   then_3     :=    PN                             :    ksi
1.14    ksi    @   EMPTY      :=    [u:ksi]CONTR                   :    'type'
1.15    ksi    @   x          :=    ---                            :    ksi
1.16    x      @   u          :=    ---                            :    EMPTY(ksi)
1.17    u      @   then_4     :=    <x>u                           :    CONTR
1.18    x      @   then_5     :=    [t:EMPTY(ksi)] then_4(t)       :    EMPTY(EMPTY(ksi))
1.19    ksi    @   PARADISE_II :=   [t:EMPTY(EMPTY(ksi))] ksi      :    'type'
```

Fig. 12 Booleans in AUT-68

```
let PEIRCE = prove
    ('((A ==> B) ==> A) ==> A',
     ASM_CASES_TAC 'A:bool' THEN ASM_REWRITE_TAC[ ]);;

let REAL_NEG_ADD = prove
  ('!x y. --(x + y) = --x + --y',
   REPEAT GEN_TAC THEN
   MATCH_MP_TAC(GEN_ALL(fst(EQ_IMP_RULE(SPEC_ALL REAL_EQ_ADD_RCANCEL))))
   THEN EXISTS_TAC 'x + y' THEN REWRITE_TAC[REAL_ADD_LINV] THEN
   ONCE_REWRITE_TAC[AC REAL_ADD_AC
                    '(a + b) + (c + d) = (a + c) + (b + d)']
   THEN REWRITE_TAC[REAL_ADD_LINV; REAL_ADD_LID]);;
```

Fig. 13 Two simple proof scripts in HOL

5.2 *Logicism*

The following contains a proof in HOL of Peirce's law (valid in classical logic)

$$((A \rightarrow B) \rightarrow A) \rightarrow A).$$

The proof is immediate because classical logic is wired in. Also a proof of $-(x + y) = (-x) + (-y)$ is displayed. As this proof requires some calculation and HOL does not have a Poincaré Principle, there are procedures outside HOL that *generate* the ephemeral proof, which is checked on the fly in all details (Fig. 13).

5.3 Platonism

In Mizar the Platonist view is followed: ZFC is built into the system. There is just a little bit of Poincaré Principle (for functions like addition) and there is no procedure to check ephemeral proofs, so formalizing in this system is comparable to writing Bourbaki style mathematics. On the other hand the system has just enough automated theorem proving that it allows a *mathematical mode*: a mathematician friendly way of interacting with the system.[64] The following example is a formulation and proof of Tarski's theorem.

THEOREM. *Let A be a set and let* $f : \mathcal{P}(A) \rightarrow \mathcal{P}(A)$ *be such that*

$$\forall X, Y \in \mathcal{P}(A).X \subseteq Y \;\Rightarrow\; f(X) \subseteq f(Y).$$

Then $\exists Z \in \mathcal{P}(A).f(Z) = Z.$

The reader will be able to follow the proof, without any knowledge of the Mizar system (the symbol 'c=' stands for $\subseteq$ and expressions like ZFMISC_1:92 refer to previously proved results) (Fig. 14).

5.4 Calculism

The system PVS has many decision methods built in. This may be seen as a rich form of the Poincaré Principle:

$$\frac{\text{true}}{A} \; A =_R \text{true}$$

where R is a particular decision method.

In the following proof script of PVS it is seen that the system can deal with reasoning with inequalities, showing that in the real numbers the following is valid:

$$\forall x, x_1, x_2, x_3 \in \mathbb{R}. \; x^2 \geq 0$$
$$\neg(x_1 * x_2 = 5 \;\&\; x_1 = 4 \;\&\; x_2 = 2)$$
$$\neg(x_1 * x_2^2 = 5 \;\&\; x_1 * x_2^2 = 5 \;\&\; x_2 = 2)$$
$$\neg(x_1^2 * x_2 = 5 \;\&\; x_1 * x_2 = 5 \;\&\; x_2 = 4)$$
$$\neg(x_1 * x_2^2 = 5 \;\&\; x_1 * x_2 = 5 \;\&\; x_2 = 2)$$
$$\neg(x_1 + 5 * x_2^2 + 20 * x_2 = 0 \;\&\; x_1 > 0 \;\&\; x_2 > 0).$$

This is something that many other systems are not good at (Fig. 15).

---

[64]In <isabelle.in.tum.de/Isar> there has been developed also mathematical mode for Isabelle. One for Coq is being considered, see [14].

```
begin
 reserve A for set;
 reserve X,Y for Subset of A;
 reserve F for Function of bool A,bool A;

theorem
 for F st for X,Y st X c= Y holds F.X c= F.Y ex X st F.X = X
proof
 let F;
 assume
A1: for X,Y st X c= Y holds F.X c= F.Y;
 consider P being Subset-Family of A such that
A2: for Y holds Y in P iff Y c= F.Y from SubFamEx;
 set X = union P;
 take X;
 for Y being set st Y in P holds Y c= F.X
 proof
  let Y be set;
  assume
A3: Y in P;
  then reconsider Y as Subset of A;
  Y c= F.Y & Y c= X by A2,A3,ZFMISC_1:92;
  hence thesis by A1;
 end;
 then
A4: X c= F.X by ZFMISC_1:94;
 then F.X c= F.(F.X) by A1;
 then F.X in P by A2;
 then F.X c= X by ZFMISC_1:92;
 hence F.X = X by A4,XBOOLE_0:def 10;
end;
```

Fig. 14 Tarski's theorem in Mizar

5.5 Intuitionism

Division with remainder states the following. Let $d \in \mathbb{N}$ with $0 < d$. Then

$$\forall n \in \mathbb{N} \exists q, r [r < d \,\&\, n = qd + r].$$

This can be proved constructively. From this proof one can automatically derive an algorithm that does the (round off) division. This is done by the (trivial[65]) operation

---

[65] The triviality is caused because of the way in an intuitionistic system a proof of a statement $\exists x . P$ is coded as a pair $\langle a, p \rangle$, where p is a proof of $P[x := a]$.

```
arith: THEORY
BEGIN

  x, x1, x2, x3, x4: VAR real

  Test1: FORMULA
    x * x >= 0

  Test2: LEMMA
    x1 * x2 = 5 and x1 = 4 and x2 = 2 IMPLIES FALSE

  Test3: LEMMA
    x1*x2 = 5 AND x1*x2*x2 = 5 AND x2 = 2 IMPLIES FALSE

  Test4: LEMMA
    x1*x2*x2 = 5 AND  x1*x2 = 5 AND  x1 = 4 IMPLIES FALSE

  Test5: LEMMA
    x1*x2*x2 = 5 AND  x1*x1*x2 = 2 AND  x2 = 2 IMPLIES FALSE

  Test6: LEMMA
    NOT ( x_1 + 5 *x_2 * x_2 + 20*x_2 = 0 AND x_1 > 0 AND x_2 > 0)

END arith
```

Fig. 15 A PVS verification based on decision methods

'choose' that assigns to a proof of an exists statement a witness. The following is part of a Coq development. The extracted code for dividing is not efficient, better extraction mechanisms and proofs will help, but it plays an explanatory service. The proof, checkable in Coq provided that a module math-mode is included. Boh have been developed by Mariusz Giero and are inspired by Mizar and Barendregt [14] (Fig. 16).

This enables one to extract an algorithm for obtaining the quotient (and the remainder as well) (Fig. 17).

Here [x:A]B stands for $\lambda x{:}A.B$ and (x:A)B for $\Pi x{:}A.B$. Notice that quotient 13 3 requires an additional argument that the divisor 3 is positive (i.e. not zero).

6 Discussion

This section really should be called: "opinions". We feel that most of the -isms are overly emphasizing a particular aspect of the mathematical endeavor. At some level, mathematics is indeed a meaningless game with symbols, and although that is not a particularly fruitful view, for the implementation of the first proof-checkers it was.

```
Lemma Euclid : (d:nat)(0<d)->
(n:nat)(EX q:nat|(([q:nat](EX r:nat|((r<d)/\n=((d[x]q)[+]r))))q)).
Proof.
Let_ d be nat. Assume
                (0 < d)                                         (A4).
LetTac P:=[n:nat]((EX q:nat|(EX r:nat|((r<d)/\n=((d[x]q)[+]r))))).
Claim
                ((n:nat)(before n P)->(P n))                    (A1).
  Let_ n be nat. Assume
                (before n P)                                    (A6).
  We need to prove (P n).
  Case 1 (n<d) (A2).
    Take zero and prove (EX r:nat|r<d/\n=d[x]zero[+]r).
    Take n and prove (n<d/\n=d[x]zero[+]n).
    As to (n<d) [by A2].
    Also (n=d[x]zero[+]n) [by times_com].
  Case 2 (n>=d) (A5).
    Claim ( (n-,d) <n).
      Have (0<n) [by A4, A5, lt_le_imp_lt].
    Hence claim done [by A4, pos_imp_mon_lt].
    Then (P (n-,d)) [by A6].
    Then consider q such that
       ([q:nat](EX r:nat|r<d/\n-,d=d[x]q[+]r)).
    Then consider r such that
       ([r:nat](r<d/\n-,d=d[x]q[+]r)) (A8).
    Take (S q) and prove (EX r:nat|r<d/\n=d[x](S q)[+]r).
    Take r and prove (r<d/\n=d[x](S q)[+]r).
    As to (r<d) [by A8].
    Now n=d[x](S q)[+]r).
    We have n = ((n-,d)[+]d) [by ge_imp_mon_plus_eq, A5] (Z1).
              _= (d[x]q[+]r[+]d) [by A8].
              _= (d[x](S q)[+]r) [by compute].
    Hence done.
  Hence (P n) [by dichotomy].
So we have proved (A1).
Finally we need to prove ((n:nat)(P n)).
Done [by cv_ind, A1].
Qed.
```

Fig. 16 Euclidean division with remainder in Coq

At some level mathematics consists of going from axioms to theorems, following logical rules. Again one forgets one aspect, the computations. Computations alone will not do, as there are many undecidable statements that are provably correct. Considering the Mathematical Universe as a fixed entity gives the working mathematician a strong drive, but one forgets that some properties require a lot of energy to find out (sometimes infinitely much, i.e. one cannot do it). Systems using formal intuitionism for computer mathematics, like Coq and Nuprl have

```
Definition choose [A:Set][P:A->Set][p:(EX a:A|(P a))]:A:=
                  Cases p of
                  (ex_intro a _) => a end.

Definition testimony [A:Set][P:A->Set][p:(EX a:A|(P a))]:
                     (P (choose ? ? p)) :=
                     (ex_rec A [a:A](P a)[q:(EX a:A|(P a))](P(choose A P q))
                     [a:A; pa:(P a)]pa p).

Definition quotient[d:nat][pos:(0<d)][n:nat] : nat :=
(choose nat ([q:nat](EX r:nat|((r<d)/\(n=(d[x]q[+]r)))))
(Euclid d pos n)).
```

Fig. 17 An extracted algorithm for obtaining the quotient

found the right middle way. On the other hand, if intuitionism is considered as a philosophy that states that mathematics only exists in the human mind, one would limit oneself to what may be called in a couple of decades 'pre-historic'[66] mathematics. True, the theories that can be fully run through in our mind constitutes *romantic mathematics*. But the expected results fully checked by computers that have been checked (by computers that have been checked)[n] by us will be *cool mathematics*. One does not want this chain to be long (as there is a possibility for erorrors at each relais). A compiled version of Coq (not satisfying the de Bruijn criterion) is needed to verify the proof of the main lemma of the four color theorem. But this compiled version has been shown correct by using the interpreted version of Coq, that does satisfy the de Bruijn criterion. So this is a useful chain of length 2.

In some sense the five small examples of a formalized proposition are somewhat disappointing: they are all similar. What seems worse, most examples can in essence be run also on the other systems. But I see this as good news. One has found the right way to implement what is needed for a foundation of mathematics. What is lacking in most systems, though, is userfriendliness.

Astronomy and biology have also had their romantic phase of going out in the fields and studying butterflies, plants and stars. The biologist at first could see everything with the naked eye. Then came the phase of the microscope. At present biologists use EM (electro-microscopy) or computers (the latter e.g. for gene sequencing). Very cool. The early astronomers could study the planets with the naked eye. Galileo started using a telescope and found the moons of Jupiter and mountains on the earth's moon. Nowadays there are sophisticated tools for observations from satellites. Again, very cool. Still, even today both biology and astronomy remain romantic, albeit in a different way. In a similar manner the coolness of Computer Mathematics will have its own romantics: human cleverness combined with computer power finding new understandable results.

---

[66]Expression comes from [120].

Added in print This paper has taken a long time to be published. An advantage is that in the meantime some impressive proof-checking results have been obtained. The following is taken from [17], Chap. 6.

6.1 Highlights

By the end of the twentieth century the technology of formalizing mathematical proofs was there, but impressive examples were missing. The situation changed dramatically during the first decade of the twenty-first century. The full formalization and computer verification of the Four Color Theorem in was achieved in Coq by Gonthier [65] (formalizing the proof in [104]); the Prime Number Theorem in Isabelle by Avigad et al. [8] (elementary proof by Selberg) and in HOL Light by Harrison [67] (the classical proof by Hadamard and de la Vallée Poussin using complex function theory). Building upon the formalization of the Four Color Theorem the Jordan Curve Theorem has been formalized by Tom Hales, who did this as one of the ingredients needed for the full formalization of his proof of the Kepler Conjecture, [66]. On September 20, 2012, the Feit-Thompson theorem, a non-trivial result in group theory stating that finite groups of odd order are solvable, has been proved by Gonthier and his research group: ⟨www.msr-inria.inria.fr/ events-news/feit-thompson-proved-in-coq⟩.

6.2 Certifying Software and Hardware

This development of high quality mathematical proof assistants was accelerated by the industrial need for reliable software and hardware. The method to certify industrial products is to fully formalize both their specification and their design and then to provide a proof that the design meets the specification.[67] This reliance on so called 'Formal Methods' had been proposed since the 1970s, but lacked to be convincing. Proofs of correctness were much more complex than the mere correctness itself. So if a human had to judge the long proofs of certification, then nothing was gained. The situation changed dramatically after the proof assistants came of age. The ARM6 processor—predecessor of the ARM7 embedded in the large majority of mobile phones, personal organizers and MP3 players—was certified, [53], by mentioned method. The seL4 operating system has been fully specified and certified, [76]. The same holds for a realistic kernel of an optimizing compiler for the C programming language, [84].

---

[67]This presupposes that the distance between the desired behaviour and the specification on the one hand, and that of the disign and realization on the other is short enough to be bridged properly.

Acknowledgements The author is indebted to Randy Pollack for his stylized rendering of Formalism, Logicism and Intuitionism as foundational views in connection with Computer Mathematics, that was the starting point of this paper. Moreover, useful information or feedback was gratefully received from Michael Beeson, Wieb Bosma, Bruno Buchberger, John Harrison, Jozef Hooman, Jesse Hughes, Bart Jacobs, Sam Owre, Randy Pollack, Bas Spitters, Wim Veldman and Freek Wiedijk.

References

1. Ackermann, W.: Zum Hilbertschen Aufbau der reellen Zahlen. Mathematische Annalen **99**, 118–133 (1928)
2. Aczel, P.: The type theoretic interpretation of constructive set theory. In: Macintyre, A., Pacholski, L., Paris, J. (eds.) Logic Colloquium'77 (Proceedings of the Conference Wrocław, 1977). Studies in Logic and the Foundations of Mathematics, vol. 96, pp. 55–66. North-Holland, Amsterdam (1978)
3. Aczel, P.: The type theoretic interpretation of constructive set theory: choice principles. In: Troelstra, A.S., van Dalen, D. (eds.) The L. E. J. Brouwer Centenary Symposium (Noordwijkerhout, 1981). Studies in Logic and the Foundations of Mathematics, vol. 110, pp. 1–40. North-Holland, Amsterdam (1982)
4. Aczel, P.: The type theoretic interpretation of constructive set theory: inductive definitions. In: Marcus, R.B., Dorn, G.J.W., Weingartner, P. (eds.) Logic, Methodology and Philosophy of Science, VII (Salzburg, 1983). Studies in Logic and the Foundations of Mathematics, vol. 114, pp. 17–49. North-Holland, Amsterdam (1986)
5. Aczel, P., Rathjen, M.: Notes on Constructive Set Theory. Technical report, Mittag Leffler Institute (2000/2001). www.ml.kva.se/preprints/meta/AczelMon_Sep_24_09_56.rdf.html
6. Andrews, P.B.: An Introduction to Mathematical Logic and Type Theory: To Truth Through Proof. Applied Logic, vol. 27. Springer, Heidelberg (2002)
7. Aristotle Organon (350 B.C.). See also classics.mit.edu/Aristotle
8. Avigad, J., Donnelly, K., Gray, D., Raff, P.: A formally verified proof of the prime number theorem. ACM Trans. Comput. Log. **9**(1–2), 1–23 (2007). http://arxiv.org/abs/cs/0509025v3
9. Bachmair, L., Ganzinger, H.: Buchberger's algorithm: a constraint-based completion procedure. In: Jouannaud, J.-P. (ed.) 1st International Conference on Constraints in Computational Logics, Munich. Lecture Notes in Computer Science, vol. 845, pp. 285–301. Springer, Heidelberg (1994)
10. Barendregt, H.P.: The Lambda calculus, its syntax and semantics. In: Studies in Logic and the Foundations of Mathematics, vol. 103, revised edn. North-Holland, Amsterdam (1984)
11. Barendregt, H.P.: Lambda calculi with types. In: Handbook of Logic in Computer Science, vol. 2, pp. 117–309. Oxford Science Publication, Oxford University Press, New York (1992)
12. Barendregt, H.P.: The impact of the lambda calculus in logic and computer science. Bull. Symb. Log. **3**(2), 181–215 (1997)
13. Barendregt, H.P., Barendsen, E.: Autarkic computations in formal proofs. J. Autom. Reason **28**(3), 321–336 (2002)
14. Barendregt, H.P.: Towards an interactive mathematical proof mode. In: Kamareddine, F. (ed.) Thirty Five Years of Automating Mathematics. Applied Logic Series, vol. 28, pp. 25–36. Springer, Heidelberg (2003)
15. Barendregt, H.P., Cohen, A.: Electronic communication of mathematics and the interaction of computer algebra systems and proof assistants. J. Symb. Comput. **32**, 3–22 (2001)
16. Barendregt, H.P., Bunder, M., Dekkers, W.: Systems of illative combinatory logic complete for first-order propositional and predicate calculus. J. Symb. Log. **58**(3), 769–788 (1993)
17. Barendregt, H.P., Dekkers, W.J.M., Statman, R.: Lambda Calculus with Types. Cambridge University Press, Cambridge (2013)

18. Barras, B.: Verification of the interface of a small proof system in coq. In: Gimenez, E., Paulin-Mohring C. (eds.) Proceedings of the 1996 Workshop on Types for Proofs and Programs, Aussois. LNCS 1512, pp. 28–45. Springer, Heidelberg (1996)

19. Barras, B.: Auto-validation d'un système de preuves avec familles inductives, Thèse de doctorat, Université Paris 7 (1999)

20. Barwise, J.: Admissible Sets and Structures: An Approach to Definability Theory. Perspectives in Mathematical Logic. Springer, Heidelberg (1975)

21. Beckmann, P.: A History of π, St. Martin's, New York (1971)

22. Beeson, M.: The mechanization of mathematics. In: Teuscher, C. (ed.) A. M. Turing Festschrift. Springer, Heidelberg (2003)

23. Bell, J.L.: A Primer of Infinitesimal Analysis. Cambridge University Press, Cambridge (1998)

24. Berger, U., Buchholz, W., Schwichtenberg, H.: Refined Program Extraction from Classical Proofs. Preprint Series: Mathematical Logic, vol. 14. Institut Mittag-Leffler, The Royal Swedish Academy of Sciences (2000, 2001)

25. Bishop, E.: Mathematics as a numerical language. In: Intuitionism and Proof Theory (Proceedings of the Summer Conference, Buffalo, 1968), pp. 53–71. North-Holland, Amsterdam (1970)

26. Bishop, E., Bridges, D.: Constructive analysis. In: Grundlehren der mathematischen wissenschaften [Fundamental Principles of Mathematical Sciences], vol. 279. Springer, Heidelberg (1985)

27. Börger, E., Erich, G., Yuri, G.: The Classical Decision Problem. Universitext. Springer, Heidelberg (2001). (Reprint of the 1997 original)

28. Brouwer, L.E.J.: The unreliability of the logical principles. 1975. In: A. Heyting (ed.) L. E. J. Brouwer: Collected Works 1: Philosophy and Foundations of Mathematics, pp. 107–111. North-Holland (1975), Amsterdam (1908)

29. Buchberger, B.: An algorithm for finding a basis for the residue class ring of a zero-dimensional polynomial ring. Dissertation, University of Innsbruck (1965)

30. Buchberger, B., Winkler, F.: Gröbner Bases and Applications. Cambridge University Press, Cambridge (1998)

31. Cantor, G.: 1885 Über die verschiedenen Standpunkte in bezug auf das Aktual Unendliche. In: In: Zermelo, E., Fraenkel, A. (eds.) Gesammelte Abhandlungen Mathematischen und Philosophischen Inhalts / Georg Cantor, pp. 370–377. Springer, Heidelberg (1932)

32. Capretta, V.: Abstraction and Computation. Dissertation, Department of Computer Science, Nijmegen University, The Netherlands, 6090 GL Nijmegen (2003)

33. Chang, C.C., Keisler, H.J.: Model Theory. Studies in Logic and the Foundations of Mathematics, vol. 73, 3rd edn. North-Holland, Amsterdam (1990)

34. Chou, S.C.: Mechanical Geometry Theorem Proving. Mathematics and its Applications, vol. 41. D. Reidel Publishing Co., Dordrecht (1988). (With a foreword by Larry Wos)

35. Church, A.: A set of postulates for the foundation of logic. Ann. Math. (second series) **33**, 346–366 (1932)

36. Church, A.: An unsolvable problem of elementary number theory. Am. J. Math. **58**, 345–363 (1936)

37. Church, A.: A formulation of the simple theory of types. J. Symb. Log. **5**, 56–68 (1940)

38. Cohen, A.M.: Communicating mathematics across the web. In: Mathematics Unlimited—2001 and Beyond, pp. 283–300. Springer, Heidelberg (2001)

39. Collins, G.E.: Quantifier elimination for real closed fields by cylindrical algebraic decomposition. In: Automata Theory and Formal Languages (Second GI Conference, Kaiserslautern). Lecture Notes in Computer Science, vol. 33, pp. 134–183. Springer, Heidelberg (1975)

40. Constable, R.L., Allen, S.F., Bromley, H.M., Cleaveland, W.R., Cremer, J.F., Harper, R.W., Howe, D.J., Knoblock, T.B., Mendler, N.P., Panangaden, P., Sasaki, J.T., Smith, S.F.: Implementing Mathematics with the Nuprl Development System. Prentice-Hall, Upper Saddle River (1986)

41. Coquand, T., Herbelin, H.: A-translation and looping combinators in pure type systems. J. Funct. Program. **4**(1), 77–88 (1994)
42. Cousineau, G., Curien, P.L., Mauny, M.: The categorical abstract machine. Sci. Comput. Program. **8**(2), 173–202 (1987)
43. Cruz-Filipe, L., Spitters, B.: Program extraction from large proof developments. In: Proceedings of TPHOLs 2003, pp. 205–220. LNCS, 2758. Springer, Heidelberg (2003)
44. Curry, H.B.: Grundlagen der kombinatorischen logic. Am. J. Math. **52**, 509–536, 789–834 (1930)
45. Curry, H.B., Feys, R.: Combinatory Logic, vol. I (with two selections by William Craig) 2nd printing. Studies in Logic and the Foundations of Mathematics. North-Holland, Amsterdam (1958)
46. van Dalen, D.: The Development of Brouwer's Intuitionism. In: Proof Theory (Roskilde, 1997), Synthese Lib. 292, pp. 117–152. Kluwer, Dordrecht (2000)
47. de Bruijn, N.G.: The mathematical language AUTOMATH, its usage, and some of its extensions. In: Symposium on Automatic Demonstration, Versailles, 1968. Lecture Notes in Mathematics, vol. 125, pp. 29–61. Springer, Heidelberg (1970)
48. Dekkers, W., Bunder, M., Barendregt, H.: Completeness of the propositions-as-types interpretation of intuitionistic logic into illative combinatory logic. J. Symb. Log. **63**(3), 869–890 (1998)
49. Dowek, G.: The stratified foundations as a theory modulo. In: Typed Lambda Calculi and Applications (Kraków, 2001). Lecture Notes in Computer Science, vol. 2044, pp. 136–150. Springer, Heidelberg (2001)
50. Dybjer, P.: A general formulation of simultaneous inductive-recursive definitions in type theory. J. Symb. Log. **65**(2), 525–549 (2000)
51. Euclid: Euclid's Elements. Green Lion Press, Santa Fe (2002). (All thirteen books complete in one volume, The Thomas L. Heath translation, Edited by Dana Densmore)
52. Feferman, S.: In the Light of Logic. Oxford University Press, Oxford (1998)
53. Fox, A.: Formal specification and verification of ARM6. In: Basin, D.A., Wolff, B. (eds.) Theorem Proving in Higher Order Logics. Lecture Notes in Computer Science, vol. 2758. Springer, Heidelberg (2003)
54. Frege, G.: Begriffsschrift und andere Aufsätze. Georg Olms Verlag, Hildesheim. Zweite Auflage. Mit E. Husserls und H. Scholz' Anmerkungen herausgegeben von Ignacio Angelelli, Nachdruck (1971)
55. Friedman, H.: Classically and intuitionistically provably recursive functions. In: Higher Set Theory (Proceedings of the Conference Mathematics, Forschungsinst., Oberwolfach, 1977). Lecture Notes in Mathematics, vol. 669, pp. 21–27. Springer, Heidelberg (1978)
56. Gabbay, M.J., Pitts, A.M.: A new approach to abstract syntax with variable binding. Form. Asp. Comput. **13**, 341–363 (2001)
57. Gauss, C.F.: Letter to H.C. Schumacher, July 12, 1831. In: Peters, C.A.F. (ed.) Briefwechsel zwischen C. F. Gauss und H. C. Schumacher, p. 269. von Esch, Altona (1862)
58. Gentzen, G.: The collected papers of Gerhard Gentzen, In: Szabo, M.E. (ed.) Studies in Logic and the Foundations of Mathematics. North-Holland, Amsterdam (1969)
59. Geuvers, H., Poll, E., Zwanenburg, J.: Safe proof checking in type theory with Y. In: Computer Science Logic, Madrid, 1999. Lecture Notes in Computer Science, vol. 1683, pp. 439–452. Springer, Heidelberg (1999)
60. Girard, J.Y., Taylor, P., Lafont, Y.: Proofs and Types. Cambridge Tracts in Theoretical Computer Science, vol. 7. Cambridge University Press, Cambridge (1989)
61. Gödel, K.: Die Vollständigkeit der Axiome des logischen Funktionalkalküls. Monatshefte für Mathematik und Physik **37**, pp. 349–360 (1930)
62. Gödel, K.: Über formal unentscheidbare Sätze der Principia Mathematica und verwandter Systeme. Monatshefte für Mathematik und Physik **38**, pp. 173–198 (1931) Translated and commented in [64]. Another English version based on course notes by Kleene and Rosser is in [63].

63. Gödel, K.: On undecidable propositions of formal mathematical systems. In: Davis, M. (ed.) The Undecidable: Basic Papers on Undecidable Propositions, Unsolvable Problems and Computable Functions, pp. 41–74. Raven Press, New York (1965). (from Mimeographed Notes on Lectures given by Gödel in 1934)

64. Gödel, K.: Collected Works, vol. I. The Clarendon Press/Oxford University Press, New York (1986). (Publications 1929–1936, Edited and with a preface by Solomon Feferman)

65. Gonthier, G.: Formal proof–the four-color theorem. Not. Am. Math. Soc. **55**(11), 1382–1393 (2008)

66. Hales, T.C.: A proof of the Kepler conjecture. Ann. Math. **162**(3), 1065–1185 (2005)

67. Harrison, J.: Formalizing an analytic proof of the prime number theorem. J. Autom. Reason. **43**(3), 243–261 (2009)

68. Hilbert, D.: Mathematical problems. Bull. Am. Math. Soc. **8**, 437–479 (1902). Earlier publications (in the original German) appeared in *Göttinger Nachrichten*, pp. 253–297 (1900), and *Archiv der Mathematik und Physik*, 3d ser., vol. 1, pp. 44–63, 213–237 (1901). See also Gray, J.J.: The Hilbert Challenge. Oxford University Press (2000)

69. Hilbert, D.: Uber das unendliche. Mathematische Annalen **95**, 161–190 (1926)

70. Howard, W.A.: The formulae-as-types notion of construction. In: To H. B. Curry: Essays on Combinatory Logic, Lambda Calculus and Formalism, pp. 480–490. Academic, London (1980)

71. Howe, D.: Reflecting the semantics of reflected proof. In: Aczel, P. (ed.) Proof Theory, pp. 229–250. Cambridge University Press, Cambridge (1992)

72. Hurkens, A.J.C.: A simplification of Girard's paradox. In: Typed Lambda Calculi and Applications, Edinburgh. Lecture Notes in Computer Science, vol. 902, pp. 266–278. Springer, Heidelberg (1995)

73. Jacobs, B.: Categorical Logic and Type Theory. Studies in Logic and the Foundations of Mathematics, vol. 141. North-Holland, Amsterdam (1999)

74. Jervell, H.R.: Thoralf Skolem: pioneer of computational logic. Nordic J. Philos. Log. **1**(2), 107–117 (1996) (electronic)

75. Kleene, S.C.: Lambda-definability and recursiveness. Duke Math. J. **2**, 340–353 (1936)

76. Klein, G., Elphinstone, K., Heiser, G., Andronick, J., Cock, D., Derrin, P., Elkaduwe, D., Engelhardt, K., Kolanski, R., Norrish, M., Sewell, T., Tuch, H., Winwood, S.: seL4: formal verification of an OS kernel, In: Matthews, J.N., Anderson, Th. (eds.) ACM Symposium on Principles of Operating Systems, Big Sky, pp. 207–220 (2009)

77. Kline, M.: Mathematical Thought from Ancient to Modern Times, vol. 1, 2nd edn. The Clarendon Press/Oxford University Press, New York (1990)

78. Klop, J.W. et al. (ed.) Term Rewrite Systems. Cambridge University Press, Cambridge (2003)

79. Knuth, D.E., Bendix, P.B.: Simple word problems in universal algebras. In: Computational Problems in Abstract Algebra (Proceedings of the Conference, Oxford, 1967), pp. 263–297. Pergamon, Oxford (1970)

80. Kreisel, G.: Proof theory and the synthesis of programs: Potential and limitations. In: Bruno Buchberger (ed.) EUROCAL'85: European Conference on Computer Algebra. Lecture Notes in Computer Science, vol. 203, pp. 136–150. Springer, Heidelberg (1985)

81. Laan, T.: The evolution of type theory in logic and mathematics. Technische Universiteit Eindhoven, Eindhoven. Dissertation, Technische Universiteit Eindhoven, Eindhoven (1997)

82. Landau, E.: Grundlagen der Analysis (das Rechnen mit ganzen, rationalen, irrationalen, komplexen Zahlen), 3rd edn. Chelsea Publishing Co., New York (1960)

83. Leibniz, G.W.: De scientia universalis seu calculo philosophico. In: Gerhardt, C.I. (ed.) Die Philosophischen Schriften von Gottfried Wilhelm Leibniz, vol. VII. Weidmann, Berlin (1875–1890). (Reprinted 1960–1961, Georg Olms Verlag, Hildesheim)

84. Leroy, X.: A formally verified compiler back-end. J. Autom. Reason. **43**(4), 363–446 (2009)

85. Letouzey, P.: A new extraction for Coq. In: Proceedings of the TYPES Conference 2002. LNCS 2626, pp. 200–219. Springer, Heidelberg (2003)

86. Makkai, M.: On structuralism in mathematics. In: Language, Logic, and Concepts, Bradford Book, pp. 43–66. MIT, Cambridge (1999)

87. Marché, C.: Normalized rewriting: an unified view of Knuth-Bendix completion and Gröbner bases computation. Prog. Comput. Sci. Appl. Log. **15**, 193–208 (1998)
88. Martin-Löf, P.: Intuitionistic type theory. In: Studies in Proof Theory. Lecture Notes 1. Bibliopolis, Naples (1984). (Notes by Giovanni Sambin)
89. Martin-Löf, P.: An intuitionistic theory of types. In: Sambin, G., Smith, J.M. (eds.) Twenty-Five Years of Constructive Type Theory. Oxford Logic Guides, vol. 36, pp. 127–172. Oxford University Press, Oxford (1998)
90. McCarthy, J.: Computer programs for checking the correctness of mathematical proofs. In: Proceedings of a Symposium in Pure Methematics, vol. V, pp. 219–227. American Mathematical Society, Providence (1962)
91. Middeldorp, A., Star, M.: A rewrite approach to polynomial ideal theory. Report CS-R9160. CWI, Amsterdam (1991)
92. Mines, R., Richman, F., Ruitenburg, W.: A Course in Constructive Algebra. Universitext. Springer, Heidelberg (1988)
93. Moerdijk, I., Palmgren, E.: Type theories, toposes and constructive set theory: predicative aspects of AST. Ann. Pure Appl. Logic **114**, 155–201 (2002)
94. Moerdijk, I., Reyes, G.E.: Models for Smooth Infinitesimal Analysis. Springer, Heidelberg (1991)
95. Nederpelt, R.P., Geuvers, J.H., de Vrijer, R.C.: Twenty-five years of Automath research. In: Selected Papers on Automath. Studies in Logic and the Foundations of Mathematics, pp. 3–54, vol. 133. North-Holland, Amsterdam (1994)
96. Newton, I.: Method of Fluxions and Infinite Series. John Nourse, London (1736). (Posthumous translation from the unpublished Latin original [1671] by J. Colson)
97. Paulin-Mohring, C.: Extracting F_ω's programs from proofs in the Calulus of Constructions. In: Sixteenth Annual ACM Symposium on Principles of Programming Languages, ACM, Austin (1989)
98. Paulin-Mohring, C.: Inductive definitions in the system Coq; rules and properties. In: Typed Lambda Calculi and Applications, Utrecht. Lecture Notes in Computer Science, vol. 664, pp. 328–345. Springer, Heidelberg (1993)
99. Péter, R.: Über den zusammenhang der verschiedenen begriffe der rekursiven funktion. Math. Annalen **110**, 612–632 (1934)
100. Péter, R.: Recursive Functions, 3rd rev. edn. (Translated from the German by István Földes). Academic, New York (1967)
101. Poincaré, H.: La Science et l'Hypothèse. Flammarion, Paris (1902)
102. Poincaré, H.: La Valeur de la Science. Flammarion, Paris (1905)
103. Pollack, R.: On extensibility of proof checkers. In: Types for Proofs and Programs. Lecture Notes in Computer Science, vol. 996, pp. 140–161. Springer, Heidelberg (1995)
104. Robertson, N., Sanders, D., Seymour, P., Thomas, R.: The four-colour theorem. J Comb. Theory Ser. B **70**(1), 2–44 (1997)
105. Robinson, A.: Non-standard analysis. In: Princeton Landmarks in Mathematics. Princeton University Press, Princeton (1996) (Reprint of the second (1974) edition, With a foreword by Wilhelmus A. J. Luxemburg)
106. Saito, M., Bernd S., Nobuki, T.: Gröbner Deformations of Hypergeometric Differential Equations. Algorithms and Computation in Mathematics, vol. 6. Springer, Heidelberg (2000)
107. Schwichtenberg, H.: Minimal logic for computable functionals. Technical report, Mathematisches Institut der Universität, München (2002)
108. Scott, D.: Constructive validity. In: Symposium on Automatic Demonstration, Versailles, 1968. Lecture Notes in Mathematics, vol. 125, pp. 237–275. Springer, Heidelberg (1970)
109. Simpson, S.G.: Partial realizations of Hilbert's Program, J. Symb. Log. **53**(2), 349–363 (1988)
110. Skolem, T.: Über ganzzahlige Lösungen einer Klasse unbestimmter Gleichungen. Norsk Matematisk Forenings skrifter **series 1**(12), 73–82 (1922)
111. Statman, R.: The typed λ-calculus is not elementary recursive. Theor. Comput. Sci. **9**(1), 73–81 (1979)

112. Sudan, G.: Sur le nombre transfini ω^ω. Bulletin mathématique de la Société Roumaine des Sciences **30**, 11–30 (1927)

113. Tarski, A.: Decision Method for Elementary Algebra and Geometry. University of California Press, Berkeley (1951)

114. Troelstra, A.S. (ed.)   Metamathematical Investigation of Intuitionistic Arithmetic and Analysis. Lecture Notes in Mathematics, vol. 344. Springer, Heidelberg (1973)

115. Troelstra, A.S., van Dalen, D.: Constructivism in Mathematics, Vol. I, II. Studies in Logic and the Foundations of Mathematics, vol. 121, 123. North-Holland, Amsterdam (1988) (An Introduction)

116. Turing, A.M.: On computable numbers, with an application to the entscheidungsproblem. Proc. Lond. Math. Soc. Ser. 2 **42**, 230–265 (1936)

117. Turing, A.M.: Computability and lambda-definability. J. Symb. Logic **2**(4), 153–163 (1937)

118. van Benthem Jutting, L.S.: Checking Landau's Grundlagen in the AUTOMATH system. Technische Hogeschool Eindhoven, Eindhoven. Doctoral dissertation, with a Dutch summary. Also in [95] (1977)

119. van Ceulen, L.: De arithmetische en geometrische fondamenten, met het ghebruyck van dien in veele verscheydene constighe questien, soo geometrice door linien, als arithmetice door irrationale ghetallen, oock door den regel Coss, ende de tafelen sinuum ghesolveert, Joost van Colster ende Jacob Marcus, Leyden (1615)

120. Zeilberger, D.:   ENCAPSULATE!, public communication. In: Cohen, A.M., Gao, X.S., Takayama, N. (eds.) Mathematical Software. First International Conference on Mathematical Software, Beijing, p. 318. World Scientific, Singapore (2002)

On the Role of Logic and Algebra in Software Engineering

Manfred Broy

Abstract Software engineering is a field of high relevance for many practical areas of advanced technology. It is essential also for a number of safety critical systems and technical infrastructures. Stimulated by the exponential growth of the power and speed of electronic hardware we observe an exponential growth in the functionality, the size, and complexity of software. In contrast to electronic hardware where we expect that everything is built based on carefully investigated theories with a deep scientific understanding, software creation still is to a large extend ad hoc. Nevertheless, due to the rising quality demands for software and the necessary improvement of productivity by advanced tools we see a growing need for a proper foundation and comprehensive theory of software engineering. In the following we outline the role that we see for logic and algebra in the emerging field of software engineering.

1 On Software Engineering

Today software is everywhere, although not always visible. More than 98 % of all programmable processors run in embedded systems. Software provides an essential part of the technical and organisational infrastructure of our world today. In software intensive business areas companies might get bankrupt within a few days or even within a few hours if their software systems are out of operation. A substantial number of people interact directly or indirectly with computers most of their working time. The reactions and man–machine-interfaces of the devices of today are essentially influenced by the concepts underlying software systems.

M. Broy (✉)
Fakultaet fuer Informatik Technische Universitaet Muenchen,
Boltzmannstr. 3 D-85748 Garching, Germany
e-mail: broy@in.tum.de; http://wwwbroy.informatik.tu-muenchen.de

P. Paule (ed.), *Mathematics, Computer Science and Logic - A Never Ending Story*,
DOI 10.1007/978-3-319-00966-7_2, © Springer International Publishing Switzerland 2013

Building software is a complex and error-prone task. Software engineering is the discipline of the development of large, powerful software systems optimising their quality, costs, and delivery schedule.

Advances in software engineering has to be based on a broad scientific and practical view and thus comprises many aspects requires a spectrum of skills. This is one reason why it is very difficult to gain measurable progress in software engineering.

In fact, researchers often tend to concentrate only on particular aspects of software engineering. However, as long as one is just concentrating on a strictly monocausal view onto software engineering such an approach is condemned to failure and will not improve the field in a substantial way. Among the many aspects the following points are of particular importance for software engineering practise.

- Economy and costs
- People, and their skills
- Development process
- Documentation – modelling & programming languages
- Application domain
- Usability – user requirements and man machine interface
- Implementation quality – correctness
- Technical quality – reliability
- Hardware and software infrastructure – performance
- Tool support – productivity
- Maintenance

In spite of the fact that software engineering has to look at and integrate many different aspects we concentrate in the following mainly on technical issues and, in particular, on the role of logic and algebra in software engineering. We argue that from a rigorous view software is, in particular, of mathematical nature and, if we want to get our hands on the foundations of software engineering, we have to understand the foundational role of logic and algebra.

Furthermore we argue that additional investigations into logic and algebra are mandatory to adopt these disciplines in an effective way for software engineering.

2   Technical Aspects in Software Engineering

We concentrate on technical aspects in software engineering and their foundations in the following. We do not look very carefully into issues of management, economy, and also not on the organisation of the development process but rather on technical and modelling issues. We discuss the technical and modelling issues nevertheless along the lines of the organisation of the development process.

In the following we go through the phases of software development as structured by the waterfall model where the idea is that each phase is to be finished before the next one is started. However, we want to emphasize that we choose these phases

only for structuring purposes. Actually the phases can be interleaved as well, as it is suggested in so-called agile, incremental or iterative software development processes.

2.1 Analysis Domain Engineering and Requirements Engineering

The most difficult and most decisive phase of software construction is not the implementation phase where the coding is done but the decision what functions and behaviours are to be realized and to be implemented and how to choose the overall structure of the development process and the software. An important, perhaps the most important step in software development is in fact the analysis and requirements phase (see [15]). In the analysis phase a first understanding of the problem domain and of the software requirements has to be achieved. Dealing with the first aspect we also speak of *domain engineering,* referring with the second aspect we speak of *requirements engineering.*

Domain engineering aims at the formalisation of a domain based on a logical theory. It depends very much on the particular domain a software engineer is looking at how much theory is initially available. In domain engineering we have to capture, analyse, understand, and document the essential notions, structures, and rules of an application domain. In essence, this means that we have to construct a data model and define on top of it the rules of the domain. In addition, we have to provide a number of behaviour models to document the behaviour and processes in the application domains. Domain engineering shows close relationships to logics. Logics is the discipline of formalizing structures and properties of subjects of discourse – which we call application domains – and ways as well as rules to reason about them.

A lot of work has to be done in informatics to formalize domains such that finally typical structures of computer science get visible that represent the relevant aspects of the domain, towards which then a detailed analysis and requirements can be targeted. Prominent examples are biology, chemistry or genetics. To do so, a domain has to be understood through the eyes of software engineers using their specific models of informatics. This is a fascinating process since informatics uses models that are quite different from the models that classical mathematics has introduced over the last two centuries.

Classical mathematics is mainly directed towards models in terms of continuous functions, integral and differential calculus. In fact, the models of informatics are much closer to logic and algebra. Informatics is based on discrete digital models. In contrast to the quantitative models of continuous mathematics, logic and algebra provide qualitative models that are able to capture the logical nature of domains in terms of data structures and functions on top of the data structures. This is one of the essential steps in making a field of application fit for computing, making a

domain accessible for a computer science treatment. In such a work we see at many places how the domains and the domain theories of application fields change and are deeply influenced by computer science modelling techniques, which are based on logic and algebra.

The relationship between the description techniques of informatics and software engineering and mathematical logic and algebra are much closer than it may seem at a first look. The description techniques of software engineering describe models that are essential the same as the models of logical formalisms. Even more, we may understand every description in software engineering, be it a table, a diagram or a text, as a logical formula (see [5]). This allows us to transfer helpful notions from logic such as consistency, completeness, soundness, and correctness to software engineering. Especially for tool support these notions are of major importance.

2.2 Sorting Out Requirements

The goal of requirements engineering is to identify and document all the requirements of the software system under construction. The first step is to prepare the requirements capture. In the requirements capture the basic requirements of software systems are identified and made precise. This activity is crucial since, if the requirements are captured incorrectly, then the resulting system will be useless, will not function in a correct way, will not address the users needs, or will be more expensive and more complex than necessary. Therefore, to find out what the actual requirements of an application system are is an important, difficult, and often painful and error prone process.

Software systems and their behaviour are highly abstract artefacts. For many users it is very difficult to imagine the functioning of software systems and all the consequences of their operation in advance. Therefore we need techniques to capture the requirements and show the consequences of imposing all kinds of requirements. Again logic can help here.

Typically requirements as they are collected from different stakeholders tend to be contradictory and inconsistent. Logic can help to define what consistency means, to analyse inconsistencies and give hints how they may be resolved.

In requirements engineering we find an interesting interplay between formal techniques from logic and algebra, techniques from domain modelling, psychology, and general techniques for structuring problems.

Roughly we can structure the process and activities of requirements engineering into the following major tasks:

- Requirements capture, identification, decision, and agreement,
- Requirements specification and documentation.

For both tasks logic and algebra can help. In the requirements capture phase we identify and discuss the choice of requirements and finally try to agree between the many people and roles involved in the particular requirements.

In the requirements specification phase we have to make the captured requirements precise. It is important to distinguish carefully between these two steps. One step makes sure that the requirements address the users' needs, the second step make sure that the requirements are formulated in a way such that the development engineers will understand them correctly to depart from them in the right direction and to show the correctness of their design and implementation.

Specifying and documenting requirements involves the description of complex properties. This is very similar to the classical techniques of mathematical logic. However, in contrast to classical logic, the number and size of the properties is often fairly huge. Therefore techniques for structuring and supporting easy understanding and reduced complexity are essential.

2.3 Design: Software and Architecture

In design we decompose a system into subsystems. This decomposition is called architecture. The architecture of a software system is essential for its manageability and maintainability. Designing architecture of a software system means structuring a software system under development into components in an appropriate way. Over the years we have learnt many lessons how to structure software systems appropriately. Modularity is only one issue here. Modularity means making sure that the components of a systems are self contained, described by proper interfaces such that we can construct and verify the architecture by just understanding the interfaces of the components. Moreover, in turn we can construct the components by only understanding their interfaces.

Although understood that way since many years due to the early work of David Parnas the theory and practice of software architecture is still, from a practical point of view, not satisfying. Many architectural descriptions are ad hoc, are not made precise enough. Architectural description languages are mostly not flexible enough to capture the complexity of sophisticated software systems.

In fact, a proper tractable theory for specifying interfaces even in object-oriented languages is not really available so far.

In software architecture, roughly speaking, we have to define a system in terms of two issues:

- A network of components,
- Its component interfaces.

The network describes the components of an architecture and how they are connected. The component interfaces describe the observable behaviour of the components as far as it is important for the functioning of the architecture.

The quality of an architecture is, of course, quite independent of the preciseness of its description. In fact, we have many problems to solve when selecting a software architecture including architecture description and justification.

First of all we have to make sure that the architecture addresses all the needs of a systems properly. This includes functional and non-functional requirements that have to be addressed by the architecture.

The role of logic and algebra in architectural design is first of all providing a proper theory of modularity such that we can prove that the decomposition of a system into components by their component interface specifications guarantees the functioning of the overall behaviour of the system. Although from a practical point of view this might be difficult to carry out, in principle, in practical situations, it is essential to have a theory ad hand available to support that.

A consequent step is to see an architecture essentially as a logical expression. As demonstrated in [6] the composition of components can be fully understood us logical composition. Then each component behaviour is captured by a logical formula, the parallel composition of components is represented by logical "and" (conjunction), the hiding of channels corresponds to existential quantification.

2.4 Stepwise Decomposition into Modules

Having described the components of an architecture each component has finally to be realized by code. To do that we decompose components further into modules. In principle, the decomposition of a component into modules is very similar to the decomposition of a system into an architecture. We can do that in a hierarchical top down decomposition as long as components are too big to be implemented by modules directly. Finally we will hopefully arrive at a size of components that can directly be realized by modules. In an object oriented terminology modules correspond to objects and classes.

Again the specification of modules has to be done in terms of logical and algebraic concepts. In contrast to software components that are more self-contained, in general, modules show lots of context dependencies. In fact, still a lot of work has to be done to find appropriate, tractable logical and algebraic foundations for the description of module interfaces.

2.5 Coding Modules

Finally after the hierarchical system decomposition into modules is finished we can go over to the module coding. In the module coding we use the module interface specifications as the starting points for the module coding. We have merely to understand the module interfaces and we do not have to look at the whole system. This reduces the complexity and the mutual dependencies of the implementation tasks considerably.

Starting from the module specifications, the implementation can be constructed in a strictly systematic top down process along the lines of structured programming as advocated by Hoare, Dijkstra, Wirth and Dahl and many more.

2.6 Code Verification

In the course of a software development, after the overall structure of a software system is fixed and the granularity of the decomposition of the components into subcomponents and modules is finished we start the implementation activities. When modules are implemented, an important issue is to verify that they are correctly realized according to their given interface specifications. This is called verification. Verification can actually be carried out in many ways. We can do it in a very logical style by logical deduction or by model checking but we can also use more pragmatic techniques such as inspection, reviews, or testing. Even for testing logic provides a firm basis and framework.

However, what is essential to keep in mind in verification is that none of the techniques really works effectively if we do not have a precise module specification. Especially testing gets into a mess if the interfaces are not specified properly. Moreover, if we are not able to verify the coding of the modules before we start to assemble them into larger building blocks the complexity of the integration of systems would be much too high and the integration process would get out of control.

A valuable technique and concept in the specification, documentation, and verification of programs are assertions. An assertion is a logical formula that uses programming constructs as logical concepts. So programming variables are used as logical variables. Functional procedures may be used as mathematical functions. Data types are used as logical types.

Assertions are logical statements about the states computer programs manipulate when executed. Assertions are helpful in many respects and are used explicitly or implicitly a lot in practice (see [14]).

2.7 System Integration and Architecture Verification

Given properly verified building blocks starting with modules we step by step compose and integrate them into larger subcomponents and in turn put together subcomponents into large components and finally assemble them into systems. To do that, of course, we have to be sure that the modules and components work together properly according to their decomposition into parts of an architecture. If the conceptual correctness of the architecture is verified with respect to the interface specifications of the components, the correctness of each component

implementation with respect to its interface specification is enough to guarantee the correctness of the overall system.

It is essential that the correctness of modules with respect to the architecture of a system can already be justified at the level of their interfaces. Therefore what we need is a theory that allows us to justify and verify the software architecture with the help of the component interface specifications independently of the verification of the correctness of the modules and the components. This is what modularity is about.

Again here the verification can be done by logical means including techniques that are applied automatically such as model checking but also by more pragmatic ones in the development process such as reviews and inspections or testing – all of them based on rigorous interface specifications.

2.8   Usability Engineering – User Centric Engineering

The most difficult and most important issue in software engineering is not only to achieve the correctness of systems according to their specifications but also the appropriateness of the specifications according to the users' needs. The problem is that this is sometimes very difficult and time consuming. A newly introduced system changes the users' view of the world. It is difficult to find out and to understand what the users really need in advance before a system is build.

However, it is also very difficult for a user to understand what the system does as it is specified by a software engineer before a prototype is available. Logical specification does not help here a lot. The only solution to this difficult problem of communication between humans today is prototyping. Prototyping means putting systems together and getting first implementations of systems by which it is demonstrated to the user what a system does and whether the system functionality is actually useful and needed. This is, of course, a very costly and very difficult issue.

The more systematic and precise the descriptions of systems are at the beginning and the better the techniques are to generate from those descriptions first demonstrators and prototypes, the easier it is to find out early about the actual needs and interests of the user. Also here logic and algebra play a major and decisive role for documentation, foundation and tool building.

Finally, logic may even be used directly for prototyping. This is the idea of logic programming and approaches that generate programs from logical specifications.

3   Discovery of a New World: The Theory of Software Engineering

As we have outlined in the last section software engineering has to look at many mutually dependable aspects of a software system but nevertheless in nearly all the steps logic and algebra can be helpful. Most importantly logic and algebra

provide a proper basis for software engineering activities, modelling and description techniques leading to a theoretical foundation.

Actually for software engineers it is always the discovery of a new world to look at the logical and the algebraic foundations of their discipline. Doing so they discover that what they do is not just putting together difficult statements and cryptic formalism and notation which may work or not according to the technical machinery but it provides them with a proper theory in which we can make sure that a particular statement is correct or not.

3.1 Mathematical, Logical, Algebraic Islands in Software Engineering

Mathematic logic and algebra provide islands of theory and foundations in the field of software engineering (see [5]). Finally they form proper grounds for building bridges. These bridges can help engineers in understanding their basic structures when describing their models, in specifying their artefacts, in finding models for what they are doing and finally in getting tool support with a justified methodology. The islands without the bridges are not very helpful, however. Only, if the theory is adjusted to the engineers' needs and integrated into their processes, it gets useful.

3.2 Denotational Semantics

Perhaps the most remarkable achievement in the foundations of programming languages and their semantics is denotational semantics (see [12, 16, 17]). According to its principles of mathematical denotations and compositionality it provides exactly what we need in software development namely a modular description tool kit of the essentials of programming languages and their abstract models such that we can put together programming languages and their behaviour in a way that we can abstract away details as much as possible. This supports abstraction and reduces complexity.

We can apply just the same principles for nowadays system description and modelling languages. Also there the concepts, notations, and treatment of denotational semantics are most appropriate because it introduces a kind of modularity that is badly needed in software engineering.

3.3 Assertions

Another very powerful technique in software specification and analysis are logical assertions. Assertions are logical formulas that are comments in the context of statements that talk about the artefacts of software in a logical style. Assertions

show a close relationship to denotational semantics. More precisely, an assertion is a logical formula that uses programming variables as logical variables and this way allows formulating logical statements about the state of a program.

Many computer scientists quite independently starting even with Alan Turing but later in particular by Floyd, Hoare, Dijkstra, Dahl and many others introduced assertions. Assertions are a powerful and elegant technique to provide precise statements about programs. For the usefulness of assertion recent papers by Tony Hoare [14] give many examples.

3.4 Abstract Data Types

Abstract data types were a major step in the foundations of data structures. Data structures are one of the most important issues of software because software needs data models. Data models are built today onto data types. Often more involved data models are provided by entity/relationship techniques and object oriented data models. Also these models can be translated into axiomatic specifications of abstract data types. All kinds of data models can be seen as descriptions of algebraic structures.

It has taken years for the computer scientists to understand that in computer science data structures are not just to be seen and used as sets but they are always to be studied in the context of their characteristic operations and relations. These operations form, together with the data structure sets, algebras in the sense of mathematics, in particular, universal heterogeneous algebras.

The properties of these algebras can again be specified nicely by algebraic axioms. This brings in a very solid foundation of data structures that can be also used and implemented by support tools. How important such concepts are can also be seen in the recent developments of UML with its Object Constraint Language OCL.

Abstract data types are in its purest form nothing but heterogeneous typed higher order equational predicate logic.

3.5 Process Algebras

Algebraic techniques apply, however, not only for axiomatization of data structures but also for control structures and for entire programs. This was the major insight derived from process algebras as they have been suggested by Robin Milner, Tony Hoare and later by many others.

The theory of process algebras shows that algebraic techniques cannot only be used to describe simple data sets but they can also be used to describe the control structure of programming languages in terms of their algebraic properties. In fact, by this, at a very high-level, data and programming languages are and can be described by algebraic laws. This leads into an algebraic theory of data, programs

and processes that fits very well together with the denotational semantics of Scott and Strachey (see [18]) because it makes explicit the algebraic theory of the notation and its model if done in a proper and appropriate way.

3.6 Temporal Logics

Another fascinating generalisation of logic in software engineering, in particular towards the needs of reactive and concurrent software systems, is temporal logics. Temporal logic has been suggested to address the issues of systems where not only a transformation of an initial state into a final state is of interest but of systems which are reactive and interactive such that many places of interaction between the system and the environment during the lifetime of a system are of major importance. Therefore it is more appropriate to describe systems that way by traces of states or actions. Temporal logic allows us to specify properties of such tracks modelling systems and to reason about them.

Temporal logic is the logic of sequences of states or actions and provides particular logical foundations for that. Pioneering work by Fred Kröger, Armir Pnueli and Zohar Manna has paved the way for an extended theory of temporal logics, which nowadays becomes more and more practical, in particular, for the specification and verification of reactive or communicating systems in connection with the model checking of temporal properties. Lamport has turned this theory much more into an engineering approach capable with systems of impressive complexity. Model checking brings together algorithmics and issues of computation with logics.

3.7 Interface Abstraction

The interface of a software system or a component is given by the patterns of interaction between the system and its environment. Interface abstraction is perhaps one of the most useful concepts of software engineering.

Without interface ideas large software systems are incomprehensible and much too complex to manage. Interface abstraction is based on core techniques from algebra and from logics. Algebra tells us how to weaken given constructions in terms of homomorphisms and quotient structures, logic tells us how to describe the interfaces in a precise way.

3.8 Model Checking

Model checking is a consequent application of techniques from automata theory and temporal logic combined with a lot of ideas to gain efficiency. In model checking the set of reachable states of a system is enumerated to check whether all of them fulfil a given property.

In principle, people have thought for long while that it would be practically impossible to check large finite state systems by considering and visiting all their states. However, according to the growing computation power of our machines and sophisticated algorithms that make it efficient to check millions and schematically even billions of states model checking today is a serious and practical technique to prove properties about programs and software systems.

4 The Role of Digital Modelling

As already pointed out one of the remarkable achievements of informatics is digital modelling technology. Digital modelling technology is different to the analog models of classical mathematics formed out of continuous functions and the mathematics of integration and differentiation.

Analog models: The modelling techniques of the nineteenth and twentieth century were continuous functions, differential and integral mathematics; we speak of quantitative models.

Digital models: The modelling techniques of digital technology the twenty-first century are discrete models of logic and algebra; we speak of qualitative models.

The foundation of digital models is logic and algebra. However, in contrast to the classical way in which logic was treated in the field of mathematical logic as a theoretical field of study, computer sciences needs an engineering usage of logic and deduction. Still, there is a long way to go to make the methods of mathematical logic practically usable in a broad sense in the field of software engineering and to exploit all the potentials of logics there.

5 On the Nature of Software Development

Many people think that the following extreme views of logic and algebra on one side and the software development on the other side cannot be integrated:

- Software development means the construction of formal/mathematical/logical models – therefore it is a formal activity – software is a mathematical object that can be analyzed by mathematical techniques, specifiable, and verifiable.

- Software development is an art and a craft; it proceeds by esoteric lore, by stepwise improvement, by trial and error – software is an artifact, immeasurable, unreliable, and unpredictable.

We, however, believe that software engineering will only make substantial progress if we manage to integrate both extreme views.

We have to understand all the logical and algebraic properties and foundations that are related with software systems and at the same time we need to have all the pragmatic understanding of the economics and the organization and the management aspects of software. This applies, in particular, also to all the technical and system aspects of software.

6 Software Development as Modeling Tasks

From conceptual point of view software development has to be understood as a modelling task. A software system is just a digital executable model. Everything in the application domain that is relevant for a software process has to be finally captured explicitly or implicitly in terms of the digital models of informatics. The better and more appropriate and more integrated these models are the easier it will be to make use of them in the software development process.

6.1 Software Development: Modeling and Description of Various Aspects of Systems

Software development can be seen as a sequence of system modeling steps. There are many aspects to model when carrying out software development:

- Application domains, their data structures, functions, laws, and processes,
- Software requirements, based on domain model comprising data models, functions, and processes,
- Software architectures (see [9]), their structures, and principles,
- Software components, their roles, interfaces, states, and behaviors,
- Programs (see [3]), their internal structure, their runs (see [1, 2, 7, 10, 11, 13]), and their implementation.

For modelling we need a family of modelling concepts and a modelling theory. For many of these aspects we have proper and helpful models at hand by now. However, it will take a while to understand the role and the depths of these models in all their details and to find a good way to relate and to integrate them and their aspects.

6.2 Modeling in Software Engineering

Systematic development of distributed interactive software systems needs basic system models. These models have to reflect the structures of distributed software systems.

Description techniques are to provide specific views and abstractions of software systems such as:

- The data view,
- The interface view,
- The distribution view,
- The process view,
- The interaction view,
- The state transition view.

The technical development of systems concentrates on working out these views that lead step by step to an implementation. Each view corresponds to a logical model and to a logical formula. The integration of these views leads to a comprehensive logical model with intricate questions of consistency.

6.3 What a Model Is in Software Engineering

The term model is used in many meanings in software engineering. People talk about data models, process models, system models, domain models and finally about meta-models. What is a model in software engineering? An annotated graph or a diagram? A collection of class names, attributes, and method names?

In our view, a model is an abstraction. In engineering, a model is always a collection of formulas, diagrams, tables expressed in some notation with a well-worked out and well-understood mathematical theory!

Hence software engineering needs mathematical modeling theories of digital systems – algebra, logic, and model theory! Logic provides a unifying frame!

We end up with a very classical view of the role and nature of models in software engineering. A model in software engineering is basically the same as a model in logic. And what is a model in logic? Basically it is an algebra, which consist of a family of sets, relations, and functions. Therefore from a very abstract formal point of view a software system is a description of a model very much as logic uses a model for explaining the semantics of logical formulas.

7 "Formal Methods" and "Software Engineering"

The term "formal method" generally refers to the application of techniques from logic and algebra in software development. Formal methods are often considered by practiconers as being inadequate, too expensive, too difficult, and "not practical".

In contrast, practical software development is often considered by formalists being ad hoc, "immature", uncontrollable, and "not an engineering discipline".

The following observation is remarkable: A high percentage (> 70 %) of large software projects fails or falls short. This reflects the state of the art today of software engineering. There are many reasons for that. Only one reason is the lack of theoretical foundations.

The practice in modeling in software engineering today is diagrams. In practice, today we find many diagrammatic methods for modeling and specification (SA, SADT, SSADM, SDL, OMT, UML, ROOM, ...). Diagrams are helpful as long as they provide easy to understand intuitive descriptions of systems and their aspects. However, often diagrams stimulate quite different associations by their various viewers, depending on their background (see [4]). Then the quite intuitive understanding is counterproductive and a dangerous source of misunderstandings.

Here a universal modeling language could help. A universal language with a well-defined meaning, well structured to provide a basis of understanding. The concept of universal modeling languages is obviously a great idea – but a closer look at languages such as UML shows, how ad hoc most of its concepts and "methods" are for the practical modeling tasks of today. At best, they reflect interesting ad hoc engineering insights and practices in engineering particular applications.

The consequence of these weaknesses and ad hoc approaches is remarkable: never have practical diagrammatic modeling been justified on the basis of a comprehensive mathematical foundation.

We only mention three disappointing but representative examples:

- UML and its statecharts dialect with its endless and fruitless discussions about its semantics.
- Behavior specification of interfaces of classes and components in object oriented modeling techniques in the presence of callbacks.
- Concurrency and cooperation: Most of the practical methods especially in object orientation seem to live in the good old days of sequential computation and do not properly address the needs of highly distributed, mobile, asynchronously cooperating software systems. The introduction of threads as it is done for instance in Java to deal with concurrency is ad hoc and error prone.

The examples show the state of the art. A much deeper understanding is required to overcome the weakness and come up with appropriate models with deep and comprehensive tool support.

8 From Logics to UML and Back to Logics

As said above practical engineering tools of today such as UML fall short in many respects. They do not provide the strong and precise modelling framework that we need so badly. They do not offer a basis of common understanding.

8.1   The Vision – An Academic View!

Today advanced practical software developers have understood that modelling is at the core of software engineering. This has led to developments like UML, the Unified Modelling Language, which claims to provide all the basics for modelling software systems. Unfortunately UML does not make much use of all the deep understanding that has been gained in models in informatics over the last 30 years in fields like denotational semantics, assertions logics, temporal logic, algebraic data types and many other research contributions.

As a result the UML approach, as technique to describe a formal model of distributed software systems the approach remains shallow, imprecise, full of contradictions, and fails to provide an integrated system model.

What we need is rather obvious:

- Foundations of modeling: A tractable scientific basis, understanding, and theory for modeling, specifying, and refining of programs, software and systems.
- Powerful models supporting refinement, levels of abstractions, multi-view modeling, domain modeling.
- Comprehensive description techniques based on these foundations.
- A family of justified engineering methods based on these foundations.
- A flexible development process model combining these methods.
- Comprehensive tool support including validation, consistency checks, verification and code generation by algorithms and methods justified by the underlying theories.

Finally we aim at modeling and its theory as an integral part of software construction as an engineering discipline.

9   A Logical and Algebraic Manifest of Software Engineering

Looking at what we have outlined so far we see that there is still a long way to go. A lot of the foundations that have been worked out over the last 30 years are understood quite well by now. Unfortunately still there is a surprising discrepancy between people that have all the theoretical understanding based on logic and algebra but do not understand their practical impacts and the people in practice which do not have the theoretical understanding and are used to their ad hoc methods.

What we need to improve the situation is a close cooperation between theoretical experts and practical experts. Only this way we can be sure that the essential practical issues are addressed with the right foundations.

A consequent foundation for the engineering of software intensive systems is obtained by logics. Each description in software engineering – be it text, diagrams,

graphics or tables is a logical statement and can be translated into a logical formula: this way logic and algebra provide unifying frames, which are not available for UML, so far.

Consequences and advantages of the semantic insights for methods and tools are obvious:

- Precise semantics (see [8, 9])
- View modelling and integration
- Notion of view consistency
- Basis for transformation, manipulation, and reasoning about systems

Logic and algebra engineering will form the basis of digital systems modelling. This will make logic practical and practice logical.

10 Outlook

We have gained a lot in the theory and foundations of software engineering over recent years. The transfer to engineering is on its way. But more theoretical work is needed. We need deeper insights, better methods and tools that can only be provided by a team of experts:

- Researchers working on the foundations in logics, algebra and mathematics,
- The designers of practical engineering methods and tools,
- The programmers and engineers in charge of practical solutions,
- Application experts modeling application domains.

Successful work, however, does not only require the interaction between these types of people – we also need *hybrid* people that have both a deep theoretical and practical understanding and therefore can build the bridges between theory and practice.

References

1. Alur, R., Holzmann, G.J., Peled, D.: An analyzer for message sequence charts. Software – Conc Tool **17**, 70–77 (1996)
2. Ben-Abdallah, H., Leue, S.: Syntactic analysis of message sequence chart specifications. Technical Report 96–12, Department of Electrical and Computer Engineering, University of Waterloo (1996)
3. Booch, G.: Object Oriented Design with Applications. Benjamin Cummings, Redwood (1991)
4. Broy, M.: Towards a formal foundation of the specification and description language SDL. Form Asp Comput **3**, 21–57 (1991)
5. Broy, M.: Mathematics of software engineering. Invited talk at MPC 95. In: Möller, B. (ed.) Mathematics of Program Construction, July 1995. Kloster Irsee, Lecture Notes of Computer Science, vol. 947, pp. 18–47. Springer, Berlin (1995)

6. Broy, M.: A logical basis for modular software and systems engineering. In: Rovan, B. (ed.) SOFSEM'98: Theory and Practice in Informatics. Lecture Notes in Computer Science, vol. 1521, pp. 19–35. Springer, Berlin/New York (1998)
7. Broy, M.: The essence of message sequence charts. Keynote speech. In: Proceedings of the International Symposium on Multimedia Software Engineering, IEEE Computer Society, pp. 42–47, 11–13 Dec 2000
8. Broy, M., Stølen, K.: Specification and Development of Interactive Systems: Focus on Streams, Interfaces, and Refinement. Springer, New York (2001)
9. Broy, M., Hofmann, C., Krüger, I., Schmidt, M.: A graphical description technique for communication in software architectures. Technische Universität München, Institut fürInformatik, TUM-I9705, February 1997 URL: http://www4.informatik.tu-muenchen.de/reports/TUM-I9705, 1997. Also in: Joint 1997 Asia Pacific Software Engineering Conference and International Computer Science Conference (APSEC'97/ICSC'97)
10. Cobben, J.M.H., Engels, A., Mauw, S., Reniers, M.A.: Formal semantics of message sequence charts. Technical Report CSR 97/19, Departement of Computing Science, Eindhoven University of Technology (1997)
11. Damm, W., Harel, D.: Breathing life into message sequence charts. Weismann Insitute Technical Report CS98-09, April 1998, revised July 1998, to appear. In: FMOODS'99, IFIP TC6/WG6.1 Third International Conference on, Formal Methods for Open Object-Based Distributed Systems, Florence, Italy, 15–18 Feb 1999
12. Dana, S.: Scott: Logic and programming languages. CACM **20**(9), 634–641 (1977)
13. Graubmann, P., Rudolph, E., Grabowski, J.: Towards a Petri Net based semantics definition for message sequence charts. In: Faergemand, O., Sarma, A. (eds.) Proceedings of the 6th SDL Forum, SDL'93: Using Objects, (1993)
14. Hoare, C.A.R.: Assertions in programming: From scientific theory to engineering practice. In: Soft-Ware 2002: Computing in an Imperfect World, First Conference, Soft-Ware 2002, Belfast. Lecture Notes in Computer Science, vol. 2311, pp. 350–351. Springer, Berlin/Heidelberg (2002)
15. Holzmann, G. J., Peled, D. A., Redberg, M. H.: Design tools for requirements engineering. Bell Labs Tech J Spec Issue Softw **2**(1), 86–95 (Spring 1997)
16. Scott, D.S.: Outline of a mathematical theory of computation. In: Proceedings of the 4th Annual Princeton Conference on Information Sciences and Systems, pp. 169–176. Princeton University, Princeton (1970)
17. Scott, D.: Data types as lattices. SIAM J **5**(3), 522–586 (1976)
18. Scott, D., Strachey, Ch.: Towards a mathematical semantics for computer languages. In: Proceedings, 21st Symposium on Computers and Automata, Polytechnic Institute of Brooklyn, pp. 19–46. Also Programming Research Group Technical Monograph PRG–6, Oxford (1971)

New Directions in the Foundations of Mathematics (2002)

Stephen Wolfram

Abstract This talk was given 10 years ago. What it says still stands, but there is now quite a bit more to say.

In the 10 years that have passed, much has been done in the exploration of the computational universe, both theoretically, and in applications, particularly in technology. (One notable result, already suspected a decade ago, is a proof of the simplest universal Turing machine.)

Also in the intervening years Wolfram|Alpha has arrived, and the notion of computational knowledge that it delivers has quite a few implications for mathematics. Today, for example, we are in the beginning stages of an ambitious project to curate all published theorems of mathematics, and automatically to generate "interesting theorems" from descriptions of mathematical structures.

1 First Part of the Talk: Introduction

First of all, happy birthday Bruno!

I'm pleased to be here. I'm sorry that the technology of this is not as perfect as it might be.

Bruno and perhaps some of the rest of you have heard for years about the science project I've been doing—which has ended up with the very large book I've written.

S. Wolfram (✉)
Wolfram Research, Champaign, USA
e-mail: s.wolfram@wolfram.com

P. Paule (ed.), *Mathematics, Computer Science and Logic - A Never Ending Story*,
DOI 10.1007/978-3-319-00966-7_3, © Springer International Publishing Switzerland 2013

What I thought I'd do today is to talk a little bit about some of what's in that book that is relevant to Bruno's interests, and perhaps some of the rest of your interests, particularly in the foundations of mathematics.

Let me first of all say a few things about what the book is about.

It's called *A New Kind of Science*, and really it has three components.

The first component is looking at simple programs as sort of a generalization of mathematics—as a new area of abstract basic science.

The second large component has to do with applying what one learns from studying simple programs to application in physics and biology and computer science and a variety of other areas.

The third component has to do with the general—one might say more philosophical implications—of the things that one learns about simple programs: in particular this thing called the Principle of Computational Equivalence, that I can try to explain a little bit about to you today.

Well, so, what's the starting point of all of this?

It comes out of thinking about natural science. If one looks at history, the exact sciences have been defined for perhaps 300 years now by the idea that one should somehow use mathematics as a way to describe the natural world. Well, if one believes that one can do theoretical science at all, it should be the case that there are some kinds of definite rules that describe the natural world. The issue is whether those rules should be ones that specifically involve just the kind of constructs of human mathematics—derivatives and integrals, numbers, things like that. Or whether perhaps there might be more general rules that are relevant to describing the natural world.

It's sort of like if one makes a computer language, for example. One has to figure out what are the appropriate primitives from which one can build all of the operations one wants to do in the computer language. The issue in natural science is what are the appropriate primitives out of which one should build theories of nature. And the traditional idea has been to use the constructs of human mathematics—mathematical equations and so on—as the elements for those primitives.

What I got interested in, 20 years ago now, was the question of whether there might be more general sets of primitives that one might be able to use in doing natural science. And sort of the obvious possibility is to think about using arbitrary sets of rules.

Well, before there were computers and programming and things like that it might not have been clear what one could possibly mean by "arbitrary sets of rules." But nowadays, with computers, one can think about arbitrary rules as being the kinds of things that can be embodied in computer programs.

So the question then is: can one use the kind of things that can be embodied in computer programs as the primitives to describe the natural world?

Well, when I first started thinking about this, I sort of assumed that the primitives that would be relevant would have to be somewhat complicated—because that's what one might guess from looking at things like mathematical equations. But I decided it would be a good idea to try looking at the simplest possible programs, and what they might do.

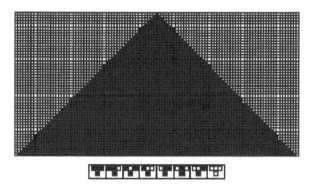

Fig. 1

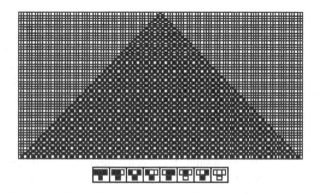

Fig. 2

Well, the particular programs I started off looking at, 20 years ago now, are things that correspond to one-dimensional cellular automata. So let me show you a little bit about what one sees in these one-dimensional cellular automata.

The way that a typical one-dimensional cellular automata works is that you have a line of cells, each either black or white. And then in a series of steps the system evolves down the page according to some definite rule that says for example that the color of a particular cell should be specified by the color of the cell above it and the cells to its left and right.

In this particular case (Fig. 1) the rule is very simple. It just says that a cell should become black whenever either it or one of its neighbors was black on the cell before. And if one then starts off with a single black cell here, one ends up getting this simple uniform black pattern.

So the question is what happens if you change the rule slightly. Here's a slightly different rule (Fig. 2). Here again, one starts off with single black cell. And one gets a simple checkerboard pattern.

So far, none of this is at all surprising. We have a very simple rule, we're starting off from a single black cell, and we're getting a very simple pattern of behavior.

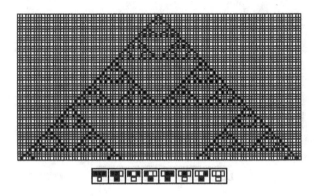

Fig. 3

But OK, let's try a different rule. Here's another rule (Fig. 3). You can specify these rules by a number, where the cases in the rule are its binary digits. This particular rule is rule number 90.[1] And with this particular rule, if you start off from a single black cell this is the pattern you get. It's not a purely repetitive pattern. So what will it end up doing? Well, here's what it ends up doing. It ends up making a nested, self-similar, fractal pattern. It's a very intricate pattern, but still, it's very regular.

And in a sense this is what one might sort of intuitively expect. One has a very simple rule, one starts off from a single black cell—surely one should get something which is a very simple regular pattern. That's certainly what I assumed would be the case, that if you sort of put in only a small rule and a small simple initial condition, that you would always get out something which is correspondingly simple.

Well, a great many years ago, I started an experiment where I systematically looked at all the possible rules.

The thing that I ended up doing was one of the simplest possible computer experiments: taking the simplest possible set of programs and just asking how they behave, starting off from a single black cell.

Well, as I mentioned, one can go through and specify by a number each of the possible simple cellular automaton rules. This one's rule number 30 (Fig. 4). And when one goes through and looks at each successive rule one eventually gets to number 30. And this is what one sees.

And this (Fig. 5) is what happens if one goes on running for longer.

Well this is a very bizarre thing, which in a sense has defined a lot of what I've done for the past nearly 20 years. It's something which at first it doesn't seem like it can be possible. Because we have a very simple rule, we're starting off from a single black cell, yet we're getting this pattern that looks extremely complicated.

---

[1]Note that $90 = 0 \cdot 2^7 + 1 \cdot 2^6 + 0 \cdot 2^5 + 1 \cdot 2^4 + 1 \cdot 2^3 + 0 \cdot 2^2 + 1 \cdot 2^1 + 0 \cdot 2^0 = [01011010]_2$.

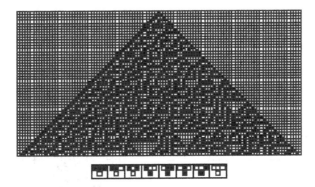

Fig. 4

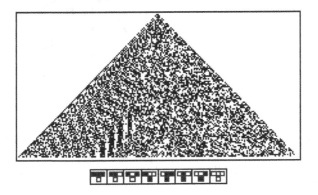

Fig. 5

There's some regularity over on the left but in general the pattern looks in many ways quite random. One might at first might think that the reason it looks so random is some problem with our visual system: that there really are regularities but we just can't see them. So one can do all sorts of elaborate mathematical and statistical analysis and so on. But one still finds that as far as one can tell, for example the center column of cells here really are completely random.

Well, it's a strange thing. Because our intuition from everyday experience and experience in building things in engineering and so on is that if we want to make something complicated, we have to start from somehow correspondingly complicated rules, and go to a lot of effort to make something complicated. But what this is saying is that no, in fact, if you allow yourself to look at arbitrary programs, many of them will produce something very complicated, just by the abstract character of what simple programs do.

And I think this particular phenomenon is the basis for a lot of important things. There has been sort of a basic issue in natural science of how nature manages to make all the complicated stuff it makes. I mean, if one is presented with two objects, one of them an artifact, one of them a natural system, it's usually a good heuristic

Fig. 6

to say that the artifact is the one that looks as if it's simpler. It seems like nature has some sort of secret that allows it in some apparently effortless way to make very complicated things. And that's something that we don't know how to do with artifacts at this point.

Well, I think that the phenomenon that we see in the case of this rule 30 cellular automaton is sort of the essential secret that nature uses to make complicated things. That even a very simple program like this can produce very complicated behavior.

The reason we don't know this is that in doing engineering it's important that the things we build are things where we can foresee what will happen. And in order to achieve this we have to constrain ourselves to use only rules where the behavior that is produced is simple enough that we can foresee what will happen.

Okay, so this rule 30 phenomenon is sort of a fundamental phenomenon. But one question is how special it is to cellular automata. Is it something that's restricted to systems that have some rigid array of cells laid out in space and so on?

Well, one can look at all different sorts of systems. Like this (Fig. 6): here's a Turing machine. Each row shows a successive state of the tape. On the left is the rule. And you can see that here the Turing machine is doing something fairly simple.

If you look at the first few steps of quite a few Turing machines you'll see only these kinds of fairly simple behavior (Fig. 7).

But if you go on, still with simple Turing machine rules, eventually you'll see this (Fig. 8).

Again very complicated behavior. The same kind of phenomenon that you see in cellular automata.

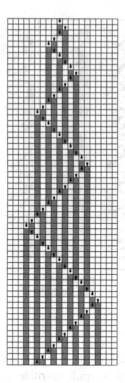

Fig. 7

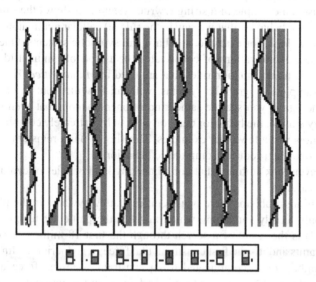

Fig. 8

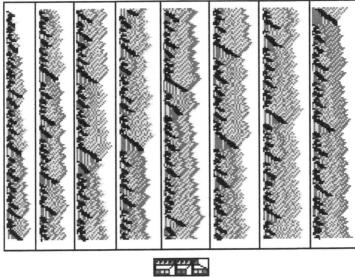

Fig. 9

OK, so what about other kinds of systems? Here's (Fig. 9) what I call a sequential substitution system. Essentially a string rewriting system where the rewrites are done sequentially. And again, one can study these, and after not very much effort one finds this—an example of a string rewrite system that shows the same kind of very complicated behavior that we saw in the rule 30 cellular automaton.

You can also look, for example, at what I call symbolic systems. One can view them either as generalizations of combinators or as sort of minimal idealizations of the symbolic transformation rules that are used in *Mathematica*. Here (Fig. 10) are some examples of symbolic systems, with rules written out in *Mathematica* notation. And again, you see the same kind of phenomenon: that even though the rules are very simple, the behavior that you get can be extremely complicated.

I'll just show you a few other examples. This (Fig. 11) is a register machine. It's the simplest one that produces complicated behavior.

You can even look at things that don't have any kind of rules of evolution—that just have constraints.

Here's a tiling problem where one specifies that everywhere in the plane, there should be only locally something which matches the particular templates shown. This (Fig. 12) is the result of a search through the first 18 million or so possible tiling constraints and it's the first constraint that forces a nonperiodic tiling to occur.

With a sightly more complicated tiling constraint, one can force a sort of a random crystal (Fig. 13) to be produced. And again, in this particular case, there's no notion of time evolution, it's just a constraint.

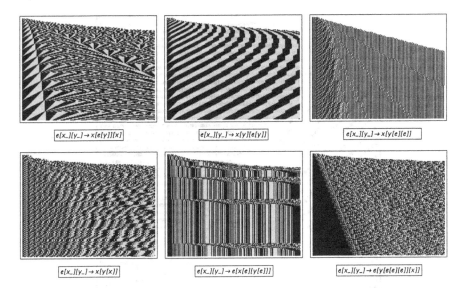

Fig. 10

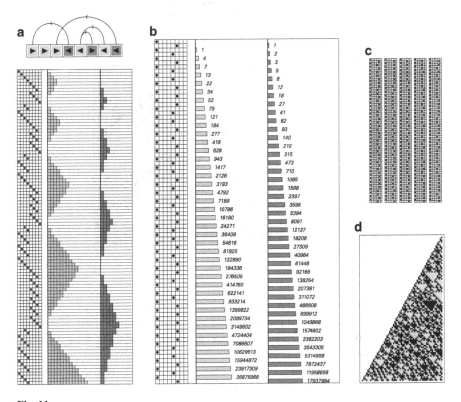

Fig. 11

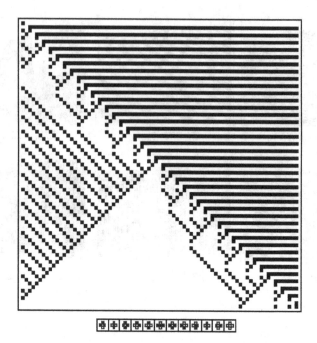

Fig. 12

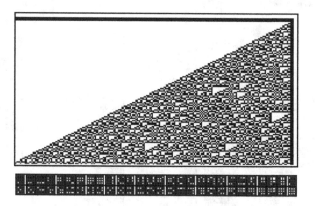

Fig. 13

One can look at systems based on numbers, as well. Here's (Fig. 14) for example what happens if you just look at powers of 3 in base 2. It's sort of a minimal version of a linear congruential random number generator, and already, it's showing very complicated behavior.

Another thing one might look at is recurrence relations. Here (Fig. 15) are some simple generalized Fibonacci-like recurrence relations. Again you see exactly the

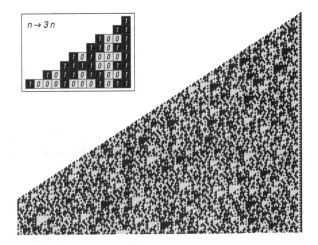

Fig. 14

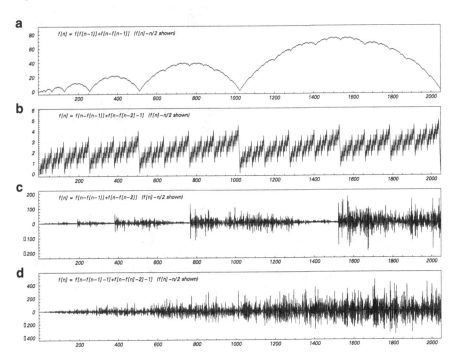

Fig. 15

same kind of phenomena: that even though the recurrence relations have a very simple form, they produce behavior that's extremely complicated.

Another case might be kind of fun for some people here. I was curious among the primitive recursive functions, what would be the first function that would show

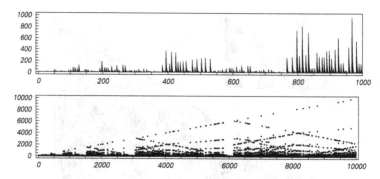

Fig. 16

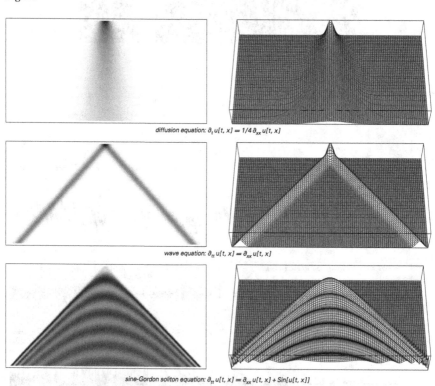

diffusion equation: $\partial_t u[t, x] = 1/4\, \partial_{xx} u[t, x]$

wave equation: $\partial_{tt} u[t, x] = \partial_{xx} u[t, x]$

sine-Gordon soliton equation: $\partial_{tt} u[t, x] = \partial_{xx} u[t, x] + \sin[u[t, x]]$

Fig. 17

complicated behavior. In *Mathematica* it's rather easy to enumerate the primitive recursive functions. And if one does that, this is the first primitive recursive function that shows complicated behavior, and here's (Fig. 16) what it does.

This kind of thing is not restricted to discrete systems. Here (Fig. 17) are some partial differential equations that are fairly commonly used. They have a fairly simple appearance.

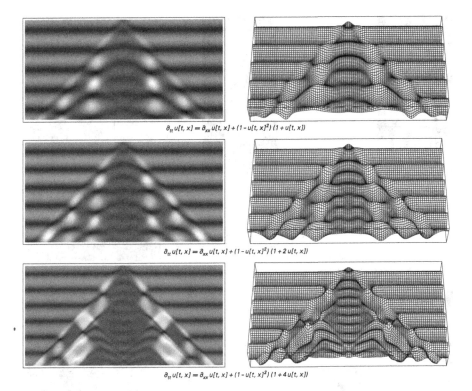

$$\partial_{tt} u[t, x] = \partial_{xx} u[t, x] + (1 - u[t, x]^2)(1 + u[t, x])$$

$$\partial_{tt} u[t, x] = \partial_{xx} u[t, x] + (1 - u[t, x]^2)(1 + 2 u[t, x])$$

$$\partial_{tt} u[t, x] = \partial_{xx} u[t, x] + (1 - u[t, x]^2)(1 + 4 u[t, x])$$

Fig. 18

But what if you search through the space of possible partial differential equations? When you think about things symbolically you can imagine just doing a search through possible symbolic expressions that represent, say, nonlinear wave equations. Well, if you do that, you eventually find this creature (Fig. 18), which is a partial differential equation that even from a simple Gaussian initial condition, produces behavior that's extremely complicated and that goes far beyond what one can analyze using traditional mathematical methods for partial differential equations.

Well, okay, so that's a little bit of an introduction to the basic phenomenon I'm talking about: that even from very simple programs one can produce these very complicated forms of behavior.

Well, one can ask how this relates to what one sees in nature. And the answer is that there are all sorts of phenomena in nature where this seems to be central to what's going on.

These (Fig. 19) are some pictures of fluid dynamics. And this is just one example where the essence of what's going on seems like it can be captured very well by thinking in terms of simple programs. In this case it seems like they can show one what's at the root of the randomness and turbulence one sees. And in general it really

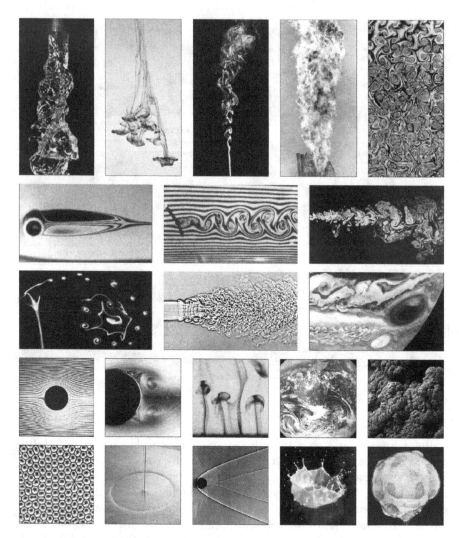

Fig. 19

seems that the primitives associated with simple programs are very effective ones for understanding the natural world.

One of the most ambitious uses of that is to try to understand things about fundamental physics.

And what I've become convinced from looking at simple programs is that it really is plausible that all of the rich and complicated things that we see in the physical universe can actually emerge from a very simple underlying program. So there might really just be a program that's, let's say, a few lines of *Mathematica* code long, that if

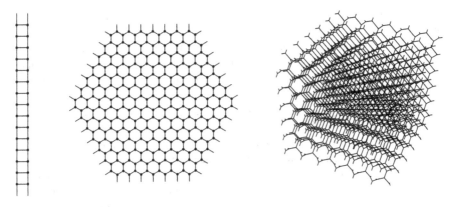

Fig. 20

one ran the program for long enough would basically reproduce in precise detail everything that one sees in the universe.

Now with the traditional view of what's going on in physics, that would seem quite implausible, but with the intuition that I've developed from looking at the behavior of simple programs, I'm very convinced that it's actually correct.

One thing that I thought I might mention is the question of what this simple program of the universe might actually be like. I think a crucial idea is that one thinks about space as being the critical element of the universe. One thinks about matter that exists in the universe as being a consequence of features of space and not really a separate thing built on top of space.

So what might space be in this kind of setup? My guess is that it'll be somehow a minimal kind of thing that has as little structure as possible. And a version which I think is a useful way to formulate it is as a trivalent network.

One can have a trivalent network that represents one-dimensional space, two-dimensional, three-dimensional and so on (Fig. 20). The only thing that one is specifying in any of these cases is the connectivity of the nodes.

Well, it turns out that one can figure out lots of things about this. It's kind of like what happens with a continuum fluid like water in a large scale: it appears continuous and smooth even though one knows that underneath there are a bunch of discrete molecules bouncing around. I think the same sort of thing happens with space: that underneath the space that we're used to there's this discrete network of nodes. Well, one can look at all sorts of properties of this.

But an important issue is to look at how time works in a system like this. This is kind of a long story. But one notion of time might be something like one sees in a cellular automaton—where every cell gets updated in parallel with every other cell at every step. That doesn't seem to be realistic for time in our universe, because if that was how it was set up, there would have to be some sort of global clock across the whole universe—which seems very implausible.

But an alternative might be that perhaps the universe works more like a Turing machine where essentially, there's just a single active cell in the universe, that moves around. Now at first that might seem like a completely crazy theory. Because it seems like the whole universe moves through time together. So how could that possibly be the result of something where there's only one place in the universe getting updated at a time?

Well, there's a very pertinent question to ask—particularly when one's dealing with remote video links like this one.[2] What if first I'm getting updated, then you're getting updated, and so on? How would we know that's what's going on? Because until I'm updated, I can't tell whether you've been updated or not.

Well, if one follows through that argument, one realizes that all that ultimately matters in terms of what we perceive in the universe is a kind of causal network of what event influences what other event. Well, here's (Fig. 21) a picture which shows how one can go from a situation where just one element of the universe gets updated at a time to a causal network, where in effect different elements of the universe are all getting updated together. Okay. So, then the question is, with this kind of setup, how do things work?

Well, let's say that one is dealing with networks in space. There's a question of how these networks should get updated. Well, one possibility is that one essentially has graph rewrite rules where essentially, any time there's a subgraph in the space network that say, looks like this, it will be replaced by a subgraph that looks like this (Fig. 22). These things go by many names, but for people here it's probably most familiar as a sort of graph rewrite system.

Now, an obvious question is what about the different possible orderings of rewrites that might occur. Well, different orderings may lead to different causal networks and so in a sense different perceived histories for the universe. Well, one might then conclude that there must be something like a whole tree of possible histories for the universe—like in many-worlds quantum mechanics—and that we populate only one branch of that tree.

That would be sort of an unfortunate theory because it would say that in order to know what actually happens in the universe, one would have to know which particular branch got picked—and knowing that goes beyond the theory one is positing for the universe. But there's another possibility, which actually relates to one of Bruno's interests, so I thought I would particularly bring it up here. The other possibility is that if the rewrite rules that one uses have a confluence property, then it turns out that the causal network that one gets is always the same, independent of what order the rewrites are done in.

Well, in the case of graphs, the condition for confluence is that there not be graphs that can overlap each other or themselves. And here (Fig. 23) are the smallest graphs of particular types that have this confluence or no-overlap property. And if you have rewrite rules for the universe that involve only these graphs, then it turns out that independent of what order you apply the rewrites into microscopically, then the

---

[2]Note: Stephen Wolfram presented this talk via a video link.

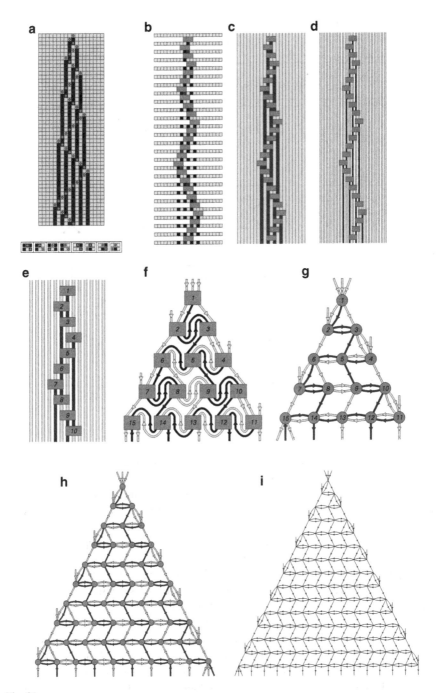

Fig. 21

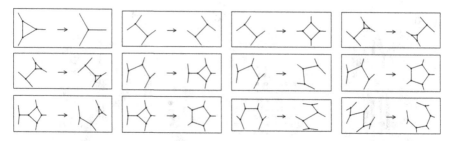

Fig. 22

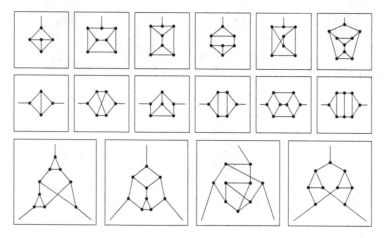

Fig. 23

confluence property implies that the causal network is always the same—and you have a property that I call causal invariance.

Well, that's a fine mathematical property, but it turns out to have a pretty important consequence. It turns out that causal invariance, subject to all sorts of detailed conditions, basically implies relativistic invariance. So, in a sense, the fact that there's causal invariance—that there's this confluence property—actually implies the validity of special relativity.

For people who are interested in physics kinds of things, what's happening is that different rewrite orders correspond to different choices of spacelike hypersurfaces. And causal invariance implies that those different choices give the same overall behavior—which is what gives relativistic invariance.

Well, it turns out that the same kind of setup also implies General Relativity. There are all sorts of very nontrivial things and details that have to be filled in. But it's a very exciting thing, that from these very simple underlying rules one seems to be able to get something that reproduces a major feature of our universe, namely gravity and General Relativity. And it's nice to be able to say here that this all sort

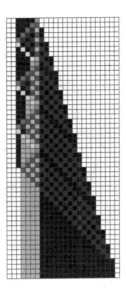

Fig. 24

of hinges on confluence properties of graph rewriting systems—something right up Bruno's alley.

Well, let me say a few more things here, and then I want to turn to foundations of mathematics.

One thing is this: that when you see the basic phenomenon of what happens in rule 30 (Fig. 4)—that even a very simple rule with very simple initial conditions can produce very complicated behavior—one can ask why does this happen, what's the fundamental reason?

Well, I think that to really understand the fundamental reason one sort of has to build a somewhat new conceptual framework—which in a sense provides a way to apply some of the deeper ideas of mathematical logic and theory of computation to natural science.

The first step is to say: "let's think about all processes in nature, in our brains, in mathematics, whatever, as computations"—where essentially there's an input to the process, an input to the computation, then it runs, and then there's output from it.

Well, the issue is then: how do we compare all these processes? We've got all these different systems and they all correspond to computations. Sometimes they will correspond to a computation that we kind of immediately know the point of. Like, here's (Fig. 24) a cellular automaton that squares any number.

You start off with say, three black cells at the top, and then at the bottom you get nine black cells—the square of the number of input cells. Or like here's (Fig. 25) a cellular automaton that computes the primes.[3]

---

[3]Consider, from left to right, the positions of the white boxes in the last row.

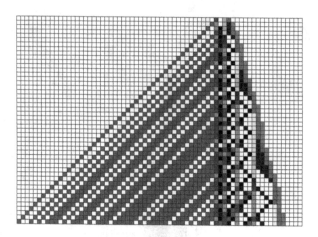

Fig. 25

These are examples of computations where in a sense we immediately know what the point of the computation is. But something like what happens in rule 30 is also a computation. It just doesn't happen to be a computation where we immediately know what the point of it is.

Well, we can ask then: how do all of these different computations compare? And our first assumption might be that different kinds of systems would do different kinds of computations. So at a practical level, to do addition, we would go and buy an adding machine, and to do exponentiation, we would go and buy an exponentiation machine. But the crucial idea, that we now know very well, is that no, that's not necessary. It's possible to have a universal machine that with appropriate input—with appropriate programming—can be set up to perform any of these functions.

And that's an idea that's now about 70 years old. And it's an idea that's been pretty important in the world, because it's the idea that makes software possible, and that really launched the whole computer evolution. But it's an idea that up till now has not been thought to be of any great relevance to natural science. And one of the things that I've tried to do is to understand what the implications of it are to natural science.

2   Second Part of the Talk: The Principle of Computational Equivalence

Well, it's a long story, but these computer experiments I've done have led to a much more aggressive version of the existence of universal computers. It's a thing that I call the Principle of Computational Equivalence. If one looks at different systems with different underlying rules, some systems will show obviously simple behavior,

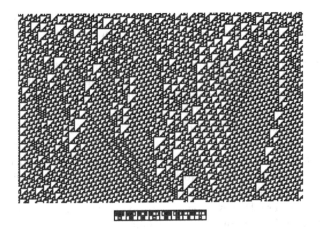

Fig. 26

let's say repetitive or nested behavior. Those systems can't exhibit sophisticated types of computation. The Principle of Computational Equivalence makes the statement that any time one sees a system whose behavior is not obviously simple, it will tend to correspond to a computation of equivalent sophistication.

So, what that means is a sharpening and a generalization of Church's thesis. Church's thesis in a modern physical form would say that for different computational processes in the universe, the maximum possible computational ability comes from a universal Turing machine or an equivalent universal system.

What the Principle of Computational Equivalence does is to go two steps beyond that. The maximum computational ability is not only achieved—that maximum is achieved for a great many systems. If one samples the space of all possible programs, many of those programs will achieve that maximum. In fact, essentially whenever one doesn't see obviously simple behavior, one will achieve that maximum level of computational ability.

Actually, the Principle of Computation Equivalence goes even further than that, and it says that not only will one achieve the maximum with appropriate initial conditions, but also, even with simple initial conditions, one will typically achieve the maximum computational ability. It isn't the case in, for example, rule 30, that one has to have some very, very elaborate initial condition to achieve behavior that corresponds to a sophisticated computation, it can be achieved with even very simple initial conditions, just like a single black cell.

Well, this Principle of Computational Equivalence is a very general thing, that at the beginning is a seed for intuition. Versions of it turn into much more precise statements. One example—the Principle of Computational Equivalence predicts that the rule 30 cellular automaton will be computationally universal.

One can look at the different simplest possible cellular automata. Here's one (Fig. 26) that I've looked at quite a bit—this is rule 110—there's its rule down there. Here, it's starting off from a random initial condition. And one sees various localized

structures getting produced that seem to interact in complicated ways. One might imagine that this is doing some kind of logic operation with all of those structures interacting with each other.

Well, it turns out, with great effort, it's possible to show that those structures can be combined in a way to emulate what is essentially a tag system, and then a Turing machine. The result shows that this little rule, this example of one of the simplest possible cellular automata, is actually universal.

This fact is of considerable significance. It might be seen to be a curious fact in the theory of computation, but it's actually a fact that's very significant for natural science. Certainly, in the past, one might have thought that in order to achieve universal computation you'd have to have something like a whole Pentium chip or some such other thing. One doesn't expect to find whole Pentium chips just lying around in nature. However, once one knows that universal computation can be achieved by a rule as simple as this, one immediately can imagine that universal computation will be very common in the natural world, and that means that all of the things we know about universal computation and simplifications can potentially apply to systems in nature.

I might mention as a footnote that this universality result implies that there's a much simpler universal Turing machine than has ever been known before. One with, well, for example, three colors and four states. Actually, I suspect that there's a Turing machine with three colors and two states that's also universal but I haven't proved that yet.

So, there's lots of universal stuff in nature. Among the consequences of this fact is a phenomenon that I call computational irreducibility, that turns out to be important in theoretical science. Normally, when one does science, one of the points is to try to find a formula for what happens in the system. One might have, for example, an idealized Earth going around the Sun, where the system does its thing, it goes around and around the Sun, but we kind of know how to shortcut that process. We don't have to trace. If we want to know where the Earth will be a million years from now we don't have to compute a million orbits of the planet around the star, we can just use some formula, plug in a number and immediately say that this is what the result will be after a million orbits.

If we think about our efforts to do prediction and compare them with the system itself, that's a case where we've essentially reduced the computational work. The system has a certain amount of computational work to figure out what it's going to do, but we've managed to shortcut that computational work and immediately skip ahead. In a sense, much of traditional theoretical science has been concerned with this idea of computational reducibility, the idea that it's possible to take what happens in nature and find a computationally reduced description of that system.

Well, in a sense, what's been done is one's made an idealization about the observer in a system. One says that the observer, computationally, is infinitely more powerful than the system that they're observing. The Principle of Computational Equivalence says that can't always be true. When the behavior of a system is complicated, the observer must be exactly computationally equivalent to the system itself.

Well, here's a consequence of that, this is a phenomenon that I call computational irreducibility. So, this is an example of a particular, this happens to be a cellular automaton, but if you look at this behavior, there's no obvious way to compress it to figure out what the outcome will be more efficiently that just following each individual step here.

And the point is that whenever there's this phenomenon of computational irreducibility, that will be what happens—we won't be able to find a compressed way to work out what the system will do much more efficiently than to just essentially watch what it does.

Well, in the limiting case that corresponds to undecidability. This is a sharper version of what one talks about in undecidability and it's a version that has the intuitive character to immediately be applied to something like natural science. One could ask about the halting problem for a cellular automaton—will it eventually die out or will it go on forever? If there's computational irreducibility, this limiting halting problem will be undecidable.

Undecidability occasionally happens in mathematics, but it is not usually deemed relevant to natural science. I've found that undecidability can be very relevant. One can ask questions, for example, in the three-body gravitation problem, whether one of the three masses will eventually escape. In chemistry, one can ask whether there will be a crystal phase that corresponds to molecules in some particular arrangement below some temperature. When one takes limiting versions of these kinds of questions in infinite time or infinite space, they can be exactly undecidable. People in physics have often been interested in finding exact solutions for the behavior of things. For example, the two-dimensional Ising model for which there's an exact solution in terms of elliptic functions and so on. The question is, what about the three-dimensional Ising model?

The kind of thing we're seeing here in terms of undecidability, computational irreducibility and so on, implies that there's going to be a limit to this exact solution business. In physical systems and other systems one will eventually get to the point where it's not possible to find exact solutions, where it's theoretically impossible.

With repetitive behavior, exact solutions are quite a simple thing in the continuous case—solved by trigonometric functions, doubly periodic functions, elliptic functions and so on. If one investigates nested patterns and fractal patterns, and one finds the functions needed to characterize them, it turns out that even with very simple such patterns, you quickly ascend in complexity. You go through binomial coefficients, then go to Gegenbauer polynomials, and then you quickly seem to ascend, even in the description of nested patterns. My guess is that all nested patterns can be in some sense described by a suitably generalized hypergeometric function, but one's already ascending to the upper limits of what's known in special function theory to even describe nested patterns. For these more complicated things, my belief is that no one will find a limited set of special functions sufficient to describe everything that's going on.

Okay, well that's a little bit about how the things that I've done relate to natural sciences. Let me turn now to discuss the implications of this for the foundations of mathematics.

Well, in a sense, mathematics is a system, where it starts from simple rules. This page basically shows all of the standard axiom systems for modern mathematics, there's logic up here, over here is geometry, category theory, set theory, etc.

These tables (Figs. 27 and 28) summarize all of the axiom systems normally used in current mathematics. From this page alone, all of mathematics as we know it grows, in some sense. One of the fundamental facts about mathematics is that it seems to be quite hard to do and, even though the axioms are simple, proofs of interesting theorems can be very long and complicated—whether that's the four-color theorem, Fermat's last theorem, or things like that. Why is it that proofs are hard to do? Why is mathematics difficult? The answer that I'm going to give at the end is that it's all related to this question of computational equivalence and the idea of computational irreducibility. To get a handle on that what one needs to do is much the same as like what I tried to do in natural science.

In natural science, one's interested in finding the appropriate primitives, to kind of capture the essential features of what's going on in systems in nature. One can ask the same questions about mathematics. What's an idealized system that captures the essence of what's going on in mathematics? Obviously, there have been various attempts to find such a thing, whether it's Whitehead and Russell's *Principia Mathematica*, whether it's category theory, whether it's any of these other kinds of things, or for example whether it's *Mathematica*. *Mathematica* is an attempt to capture the essence of what one needs to do mathematics with symbolic programming and transformation rules. One obvious question—what is the simplest idealized model that will capture the essential features of the mathematical process? I have very practical reasons to care about the answer to that question.

But so, what's essential in mathematics? Well, something that actually Bruno has long emphasized to me is that proofs are an essential feature of mathematics as it's practiced. In the kind of things that I've studied, let's say cellular automata, what one's mostly interested in is to take the system, run it and see what it does. There's an issue in mathematics of doing proofs of things, it's typically important that for proofs where there's an infinite set of possibilities, you have to represent them symbolically, and then make some general statement about them.

Well, this is an example of a typical proof in mathematics (Fig. 29). In propositional logic, this is an equational version of an axiom system. It says, if you have a propositional logic statement that looks like this, it can be rewritten like this. These are then the steps of a proof, this particular proof says that "p nand q" is equal to "q nand p", and it does that by applying these axioms in a sequence of steps to go from the input to the output.

Okay, so that's a simple proof in mathematics. One can idealize that even further to a string rewrite system. Here (Fig. 30) are the axioms in some string rewrite system and these are proofs of a few theorems that say that, for example, that this string is equivalent to that string. Well now, given one of these rewrite systems, what is the set of all provable theorems? For this rewrite system down here, for example, what string can be generated from another string? In a sense, every path through this diagram corresponds to the proof of some theorem.

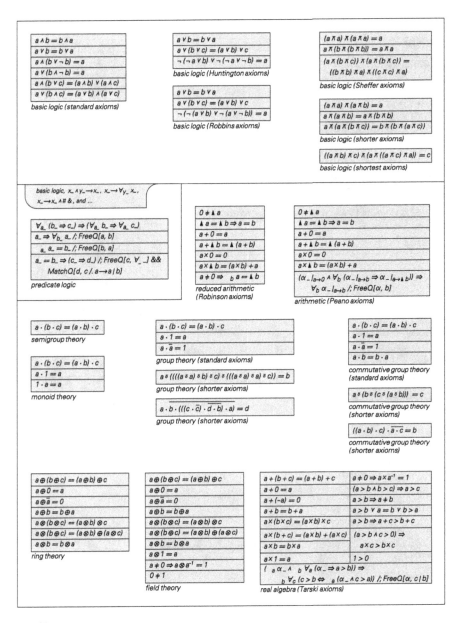

Fig. 27

So, the critical fact then is that there are theorems that involve the equivalence of short strings, but where the shortest proof of that theorem is very long. It involves very many steps to go from one string to the other and that's this phenomenon of computational irreducibility. Essentially this phenomenon says that there will be

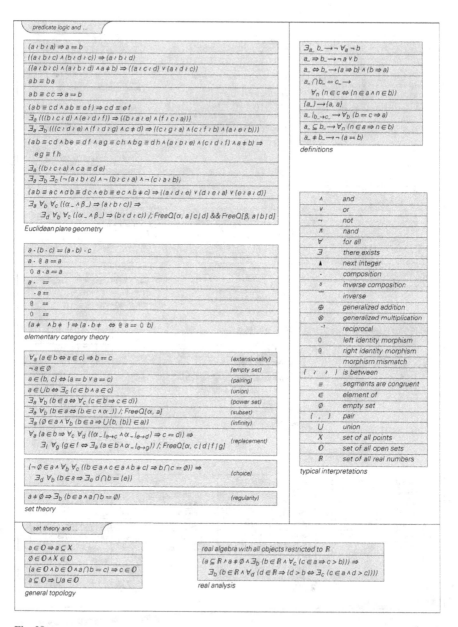

Fig. 28

short strings for which the shortest way to get from one string to another involves an irreducibly long computation, and that's why long proofs occur in mathematics.

Using this setup as an idealization of mathematics, there are several more steps one needs to take. One that's important is to introduce a notion of truth and falsity.

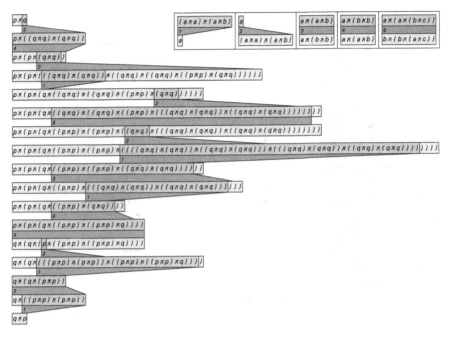

Fig. 29

Right now we're just dealing with strings, here. One can introduce the notion of truth and falsity and the idea of negation on strings. One of the things that comes out—there is an empirical version of completeness and incompleteness in Gödel's theorem. A system that doesn't generate enough strings is likely an incomplete system. A system that generates too many strings is likely an inconsistent system. In a very straightforward and visual way, one can look at the empirical version of Gödel's theorem. If one looks at the set of all possible string rewrite systems or all possible axiom systems, how many of them are very inconsistent, almost consistent, almost complete, or very incomplete? How many are actually complete and consistent? From all this, there's a distribution of possible axiom systems and their properties.

Well, one question then is ... let me talk a little bit about Gödel's theorem, and the question of whether Gödel's theorem is actually relevant to everyday life. I mean, in the original version of Gödel's theorem, the statement "This statement is unprovable" is a statement which Gödel, with great effort, managed to show really was a statement in arithmetic. Essentially, he did that by showing that arithmetic in modern terms is universal. So one can code that statement. But it's a sufficiently complicated, metamathematical statement that one wouldn't expect that it would show up within everyday practice of mathematics, and indeed it hasn't.

It's been known for a number of years that if one looks at integer solutions of diophantine equations that there are unprovable statements that can be made about

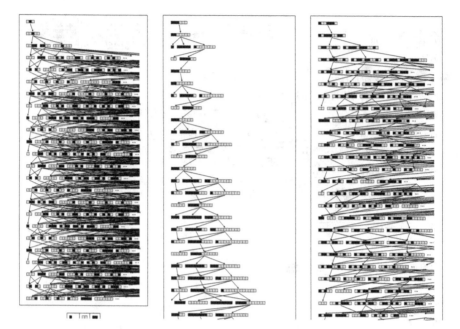

Fig. 30

$$(-3x_6 + x_7 + x_8)^2 + (2^{1+x_3 (1+x_1+2x_3)} x_2 - 2x_4 - x_{10} + x_{11})^2 + (-2x_8 - x_9 + x_{10} + x_{11})^2 + (1 - 2^{(1+x_9)(x_1+2x_3)} + x_4 + x_{12})^2 +$$
$$(1 - 2^{x_1} + x_2 + x_{13})^2 + (1 - 2^{x_1} + x_5 + x_{14})^2 + (-x_4 + 2^{x_3} x_5 + 2^{x_1+2x_3} x_6 + 2^{x_1+x_3} x_{15} + x_{16})^2 + (1 - 2^{x_3} + x_{15} + x_{17})^2 +$$
$$(1 - 2^{x_3} + x_{16} + x_{18})^2 + (-x_6 - 2x_7 + x_9 + x_{19})^2 + (-(2 + 2^{x_6})^{x_6} + (1 + 2^{x_6})^{x_7} (1 + 2x_{20} + (1 + 2^{x_6}) x_{21}) + x_{22})^2 + (1 - (1 + 2^{x_6})^{x_7} + x_{22} + x_{23})^2 +$$
$$(1 - 2^{x_6} + 2x_{20} + x_{24})^2 + (-(2 + 4^{x_6})^{2x_6} + (1 + 4^{x_6})^{x_7} (1 + 2x_{25} + (1 + 4^{x_6}) x_{26}) + x_{27})^2 + (1 - (1 + 4^{x_6})^{x_7} + x_{27} + x_{28})^2 +$$
$$(1 - 4^{x_6} + 2x_{25} + x_{29})^2 + (-(2 + 2^{x_6})^{x_6} + (1 + 2^{x_6})^{x_6} (1 + 2x_{30} + (1 + 2^{x_6}) x_{31}) + x_{32})^2 + (1 - (1 + 2^{x_6})^{x_6} + x_{32} + x_{33})^2 +$$
$$(1 - 2^{x_6} + 2x_{30} + x_{34})^2 + (-(2 + 2^{x_6})^{x_6} + (1 + 2^{x_6})^{2x_6} (1 + 2x_{35} + (1 + 2^{x_6}) x_{36}) + x_{37})^2 + (1 - (1 + 2^{x_6})^{2x_6} + x_{37} + x_{38})^2 +$$
$$(1 - 2^{x_6} + 2x_{35} + x_{39})^2 + (-(2 + 2^{x_6})^{x_6} + (1 + 2^{x_6})^{x_9} (1 + 2x_{40} + (1 + 2^{x_6}) x_{41}) + x_{42})^2 + (1 - (1 + 2^{x_6})^{x_9} + x_{42} + x_{43})^2 +$$
$$(1 - 2^{x_6} + 2x_{40} + x_{44})^2 + (-(2 + 4^{x_7})^{2x_7} + (1 + 4^{x_7})^{x_9} (1 + 2x_{45} + (1 + 4^{x_7}) x_{46}) + x_{47})^2 + (1 - (1 + 4^{x_7})^{x_9} + x_{47} + x_{48})^2 +$$
$$(1 - 4^{x_7} + 2x_{45} + x_{49})^2 + (-(2 + 2^{x_{19}})^{x_{19}} + (1 + 2^{x_{19}})^{x_6} (1 + 2x_{50} + (1 + 2^{x_{19}}) x_{51}) + x_{52})^2 + (1 - (1 + 2^{x_{19}})^{x_6} + x_{52} + x_{53})^2 +$$
$$(1 - 2^{x_{19}} + 2x_{50} + x_{54})^2 + (-(2 + 2^{x_{19}})^{x_{19}} + (1 + 2^{x_{19}})^{2x_7} (1 + 2x_{55} + (1 + 2^{x_{19}}) x_{56}) + x_{57})^2 + (1 - (1 + 2^{x_{19}})^{2x_7} + x_{57} + x_{58})^2 +$$
$$(1 - 2^{x_{19}} + 2x_{55} + x_{59})^2 + (-(2 + 2^{x_9})^{x_9} + (1 + 2^{x_9})^{x_{10}} (1 + 2x_{60} + (1 + 2^{x_9}) x_{61}) + x_{62})^2 + (1 - (1 + 2^{x_9})^{x_{10}} + x_{62} + x_{63})^2 + (1 - 2^{x_9} + 2x_{60} + x_{64})^2 +$$
$$(-(2 + 4^{x_8})^{2x_8} + (1 + 4^{x_8})^{x_{10}} (1 + 2x_{65} + (1 + 4^{x_8}) x_{66}) + x_{67})^2 + (1 - (1 + 4^{x_8})^{x_{10}} + x_{67} + x_{68})^2 + (1 - 4^{x_8} + 2x_{65} + x_{69})^2 +$$
$$(-(2 + 2^{x_{11}})^{x_{11}} + (1 + 2^{x_{11}})^{x_9} (1 + 2x_{70} + (1 + 2^{x_{11}}) x_{71}) + x_{72})^2 + (1 - (1 + 2^{x_{11}})^{x_9} + x_{72} + x_{73})^2 + (1 - 2^{x_{11}} + 2x_{70} + x_{74})^2 +$$
$$(-(2 + 2^{x_{11}})^{x_{11}} + (1 + 2^{x_{11}})^{2x_8} (1 + 2x_{75} + (1 + 2^{x_{11}}) x_{76}) + x_{77})^2 + (1 - (1 + 2^{x_{11}})^{x_8} + x_{77} + x_{78})^2 + (1 - 2^{x_{11}} + 2x_{75} + x_{79})^2 = 0$$

Fig. 31

diophantine equations. Here (Fig. 31) is an exponential diophantine equation that I constructed based on the rule 110 cellular automaton. A statement such as "this diophantine equation has no solutions for certain variables settings" is something that one can explicitly see is undecidable. It's a concrete instance of Gödel's theorem, but this equation is complicated enough that you wouldn't expect to kind of find it lying around, to find it in practical mathematics.

$2x+3y=1$	Ł
$2x+3y=2$	Ł
$2x+3y=3$	Ł
$2x+3y=4$	Ł
$2x+3y=5$	x = 1, y = 1
$2x+3y=6$	Ł
$2x+3y=7$	x = 2, y = 1
$2x+3y=8$	x = 1, y = 2
$2x+3y=9$	x = 3, y = 1
$2x+3y=10$	x = 2, y = 2
$2x+3y=11$	x = 1, y = 3
$2x+3y=12$	x = 3, y = 2
$2x+3y=13$	x = 2, y = 3
$2x+3y=14$	x = 1, y = 4
$2x+3y=15$	x = 3, y = 3
$x^2=y^2+1$	Ł
$x^2=y^2+2$	Ł
$x^2=y^2+3$	x = 2, y = 1
$x^2=y^2+4$	Ł
$x^2=y^2+5$	x = 3, y = 2
$x^2=y^2+6$	Ł
$x^2=y^2+7$	x = 4, y = 3
$x^2=y^2+8$	x = 3, y = 1
$x^2=y^2+9$	x = 5, y = 4
$x^2=y^2+10$	Ł
$x^2=y^2+11$	x = 6, y = 5
$x^2=y^2+12$	x = 4, y = 2
$x^2=y^2+13$	x = 7, y = 6
$x^2=y^2+14$	Ł
$x^2=y^2+15$	x = 4, y = 1
$x^2=y^2+16$	x = 5, y = 3
$x^2=y^2+1$	Ł
$x^2=2y^2+1$	x = 3, y = 2
$x^2=3y^2+1$	x = 2, y = 1
$x^2=4y^2+1$	Ł
$x^2=5y^2+1$	x = 9, y = 4
$x^2=6y^2+1$	x = 5, y = 2
$x^2=7y^2+1$	x = 8, y = 3
$x^2=8y^2+1$	x = 3, y = 1
$x^2=9y^2+1$	Ł
$x^2=10y^2+1$	x = 19, y = 6
$x^2=11y^2+1$	x = 10, y = 3
$x^2=12y^2+1$	x = 7, y = 2
$x^2=13y^2+1$	x = 649, y = 180
$x^2=14y^2+1$	x = 15, y = 4
$x^2=15y^2+1$	x = 4, y = 1
$x^2=16y^2+1$	Ł
$x^2=17y^2+1$	x = 33, y = 8
$x^2=18y^2+1$	x = 17, y = 4
$x^2=19y^2+1$	x = 170, y = 39
$x^2=20y^2+1$	x = 9, y = 2

$x^2=y^3-20$	x = 14, y = 6
$x^2=y^3-19$	x = 18, y = 7
$x^2=y^3-18$	x = 3, y = 3
$x^2=y^3-17$	Ł
$x^2=y^3-16$	Ł
$x^2=y^3-15$	x = 7, y = 4
$x^2=y^3-14$	Ł
$x^2=y^3-13$	x = 70, y = 17
$x^2=y^3-12$	Ł
$x^2=y^3-11$	x = 4, y = 3
$x^2=y^3-10$	Ł
$x^2=y^3-9$	Ł
$x^2=y^3-8$	Ł
$x^2=y^3-7$	x = 1, y = 2
$x^2=y^3-6$	Ł
$x^2=y^3-5$	Ł
$x^2=y^3-4$	x = 2, y = 2
$x^2=y^3-3$	Ł
$x^2=y^3-2$	x = 5, y = 3
$x^2=y^3-1$	Ł
$x^2=y^3$	x = 1, y = 1
$x^2=y^3+1$	x = 3, y = 2
$x^2=y^3+2$	Ł
$x^2=y^3+3$	x = 2, y = 1
$x^2=y^3+4$	Ł
$x^2=y^3+5$	Ł
$x^2=y^3+6$	Ł
$x^2=y^3+7$	Ł
$x^2=y^3+8$	x = 3, y = 1
$x^2=y^3+9$	x = 6, y = 3
$x^2=y^3+10$	Ł
$x^2=y^3+11$	Ł
$x^2=y^3+12$	x = 47, y = 13
$x^2=y^3+13$	Ł
$x^2=y^3+14$	Ł
$x^2=y^3+15$	x = 4, y = 1
$x^2=y^3+16$	Ł
$x^2=y^3+17$	x = 5, y = 2
$x^2=y^3+18$	x = 19, y = 7
$x^2=y^3+19$	x = 12, y = 5
$x^2=y^3+20$	Ł
$x^2=y^3+1$	x = 3, y = 2
$x^2=2y^3+1$	Ł
$x^2=3y^3+1$	x = 2, y = 1
$x^2=4y^3+1$	Ł
$x^2=5y^3+1$	Ł
$x^2=6y^3+1$	x = 7, y = 2
$x^2=7y^3+1$	Ł
$x^2=8y^3+1$	x = 3, y = 1
$x^2=9y^3+1$	Ł
$x^2=10y^3+1$	x = 9, y = 2

$x^3=y^4-20xy-1$	x = 10, y = 7
$x^3=y^4-19xy-1$	x = 3, y = 4
$x^3=y^4-18xy-1$	x = 75, y = 26
$x^3=y^4-17xy-1$	
$x^3=y^4-16xy-1$	
$x^3=y^4-15xy-1$	x = 624, y = 125
$x^3=y^4-14xy-1$	
$x^3=y^4-13xy-1$	
$x^3=y^4-12xy-1$	x = 3, y = 2
$x^3=y^4-11xy-1$	
$x^3=y^4-10xy-1$	
$x^3=y^4-9xy-1$	x = 80, y = 27
$x^3=y^4-8xy-1$	x = 12, y = 7
$x^3=y^4-7xy-1$	x = 1, y = 2
$x^3=y^4-6xy-1$	x = 15, y = 8
$x^3=y^4-5xy-1$	
$x^3=y^4-4xy-1$	x = 30, y = 13
$x^3=y^4-3xy-1$	
$x^3=y^4-2xy-1$	
$x^3=y^4-xy-1$	
$x^3=y^4-1$	Ł
$x^3=y^4+xy-1$	x = 1, y = 1
$x^3=y^4+2xy-1$	x = 3, y = 2
$x^3=y^4+3xy-1$	x = 5, y = 3
$x^3=y^4+4xy-1$	x = 2, y = 1
$x^3=y^4+5xy-1$	
$x^3=y^4+6xy-1$	
$x^3=y^4+7xy-1$	
$x^3=y^4+8xy-1$	x = 20, y = 9
$x^3=y^4+9xy-1$	x = 3, y = 1
$x^3=y^4+10xy-1$	
$x^3=y^4+11xy-1$	x = 5, y = 2
$x^3=y^4+12xy-1$	
$x^3=y^4+13xy-1$	
$x^3=y^4+14xy-1$	
$x^3=y^4+15xy-1$	
$x^3=y^4+16xy-1$	x = 4, y = 1
$x^3=y^4+17xy-1$	
$x^3=y^4+18xy-1$	x = 8, y = 3
$x^3=y^4+19xy-1$	
$x^3=y^4+20xy-1$	
$x^2=y^5+3$	x = 2, y = 1
$x^2=y^5+y+3$	x = 2537, y = 23
$x^2=y^5+2y+3$	
$x^2=y^5+3y+3$	
$x^2=y^5+4y+3$	
$x^2=y^5+5y+3$	x = 3, y = 1
$x^2=y^5+6y+3$	
$x^2=y^5+7y+3$	x = 7, y = 2
$x^2=y^5+8y+3$	
$x^2=y^5+9y+3$	

$x^3+y^3=z^2+1$	x = 1, y = 1, z = 1
$x^3+y^3=z^2+2$	x = 107, y = 232, z = 3703
$x^3+y^3=z^2+3$	x = 1, y = 3, z = 5
$x^3+y^3=z^2+4$	x = 5, y = 12, z = 43
$x^3+y^3=z^2+5$	x = 1, y = 2, z = 2
$x^3+y^3=z^2+6$	x = 7, y = 24, z = 119
$x^3+y^3=z^2+7$	x = 2, y = 2, z = 3
$x^3+y^3=z^2+8$	x = 1, y = 2, z = 1
$x^3+y^3=z^2+9$	x = 3, y = 7, z = 19
$x^3+y^3=z^2+10$	x = 2, y = 3, z = 5
$x^3+y^3=z^3-20$	x = 107, y = 137, z = 156
$x^3+y^3=z^3-19$	x = 14, y = 16, z = 19
$x^3+y^3=z^3-18$	x = 1, y = 2, z = 3
$x^3+y^3=z^3-17$	x = 103, y = 111, z = 135
$x^3+y^3=z^3-16$	x = 10, y = 12, z = 14
$x^3+y^3=z^3-15$	x = 262, y = 265, z = 332
$x^3+y^3=z^3-14$	Ł
$x^3+y^3=z^3-13$	Ł
$x^3+y^3=z^3-12$	x = 5725013, y = 9019406, z = 9730705
$x^3+y^3=z^3-11$	x = 2, y = 2, z = 3
$x^3+y^3=z^3-10$	x = 3, y = 3, z = 4
$x^3+y^3=z^3-9$	x = 52, y = 216, z = 217
$x^3+y^3=z^3-8$	x = 16, y = 12, z = 18
$x^3+y^3=z^3-7$	x = 605809, y = 680316, z = 812918
$x^3+y^3=z^3-6$	x = 1, y = 1, z = 2
$x^3+y^3=z^3-5$	Ł
$x^3+y^3=z^3-4$	Ł
$x^3+y^3=z^3-3$	
$x^3+y^3=z^3-2$	x = 5, y = 6, z = 7
$x^3+y^3=z^3-1$	x = 6, y = 8, z = 9
$x^3+y^3=z^3$	Ł
$x^3+y^3=z^3+1$	x = 1, y = 2, z = 2
$x^3+y^3=z^3+2$	x = 1214928, y = 3480205, z = 3528875
$x^3+y^3=z^3+3$	x = 4, y = 4, z = 5
$x^3+y^3=z^3+4$	Ł
$x^3+y^3=z^3+5$	Ł
$x^3+y^3=z^3+6$	x = 10529, y = 60248, z = 60355
$x^3+y^3=z^3+7$	x = 32, y = 104, z = 105
$x^3+y^3=z^3+8$	x = 1, y = 2, z = 1
$x^3+y^3=z^3+9$	x = 2097, y = 11305, z = 11329
$x^3+y^3=z^3+10$	x = 130, y = 141, z = 171
$x^3+y^3=z^3+11$	x = 297, y = 619, z = 641
$x^3+y^3=z^3+12$	x = 7, y = 10, z = 11
$x^3+y^3=z^3+13$	Ł
$x^3+y^3=z^3+14$	Ł
$x^3+y^3=z^3+15$	x = 2, y = 2, z = 1
$x^3+y^3=z^3+16$	x = 2429856, y = 6960410, z = 7057760
$x^3+y^3=z^3+17$	x = 25, y = 50, z = 52
$x^3+y^3=z^3+18$	x = 94, y = 101, z = 123
$x^3+y^3=z^3+19$	x = 26, y = 76, z = 77
$x^3+y^3=z^3+20$	x = 1, y = 3, z = 2

Fig. 32

If one looks at diophantine equations that have actually shown up, here (Fig. 32) are some examples that with great effort I found solutions to, in some cases. If one looks at the history of diophantine equations, linear diophantine equations were cracked in antiquity, quadratic ones by Gauss around 1800, and then every 50 years or so, there seems to be another class of diophantine equations that get cracked. The question is whether that process will go on—if it's going to be the case that all reasonable-sized diophantine equations are eventually going to be cracked—or

is there going to come a point when one hits undecidability, and where one can't go on with this process of cracking diophantine equations. My guess is that many of the currently unsolved questions in number theory will turn out to be undecidable, and that there's actually a boundary of undecidability, universality, and so on, which is very close at hand, and that's something that comes out from what I've studied with simple programs and so on.

Well, if it's the case that undecidability is close at hand, why hasn't it shown up more in practical mathematics? There are a couple of possible explanations for that. One is that mathematics has arranged itself to avoid undecidability, the other is that there is something essential about mathematics, which is more special than the general string rewrite system, that somehow manages to avoid the clutches of undecidability.

My own guess starts by looking at history. Mathematics, as it is practiced today, emerged from the generalizations of arithmetic and geometry that were studied in ancient Babylon. They're done by saying "what's the most general set of things that preserves such and such a theorem?" and by making generalizations in that way, inevitably one finds things where the theorems can still be proved.

Mathematics has evolved to retain generalizations that allow theorems and proofs. Have the best generalizations been picked? This picture (Fig. 33) is an example of an ultimately dessicated form of mathematics, where possible axiom systems go down the left, possible theorems go across the top, and there's a black dot whenever a particular theorem is true for a particular axiom system.

This picture actually wasn't all that easy to make, it involved all sorts of automated theorem proving to make this picture. But what this means is that every row of this picture is effectively a new field of mathematics with a certain distribution of theorems. And one question is, is there something special about the particular fields of mathematics that we've studied so far?

Where do fields of mathematics occur in the lexigraphic order of possible axiom systems? For example, I was curious about the case of logic. Well, the normal axiom systems for logic—here (Fig. 34) are some typical examples—they're fairly complicated, and in the most obvious encoding this would be the quadrillionth possible axiom system that one would write down. Is this the simplest representation of logic, or does it actually occur earlier? One thing to notice is that one can replace the multiple operators here by the single Sheffer stroke operator—a single nand operator [Sheffer axioms in Fig. 34].

So one thing I was curious about is—I had the intuition that there should be a simpler axiom system for logic, and so I started just doing a search among possible axiom systems and trying for each axiom system to prove whether it was an axiom system for logic, and it turns out that the 50,000th axiom system that I looked at was this one here. With considerable effort and automated theorem-proving technology and so on, it turned out to be possible to find a proof that this single axiom [shortest axioms in Fig. 34] is actually an axiom system for propositional logic. I know from the way the search was done that it's actually the simplest axiom system for propositional logic.

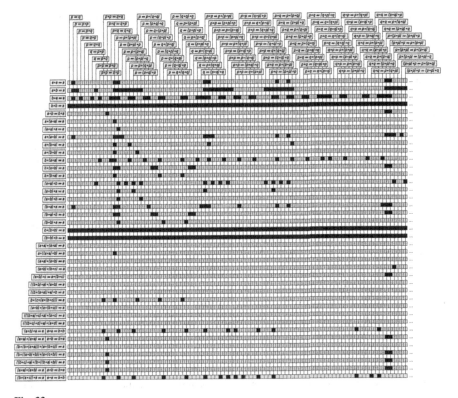

Fig. 33

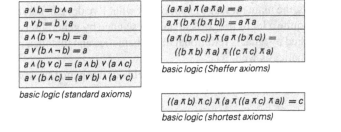

$a \wedge b = b \wedge a$
$a \vee b = b \vee a$
$a \wedge (b \vee \neg b) = a$
$a \vee (b \wedge \neg b) = a$
$a \wedge (b \vee c) = (a \wedge b) \vee (a \wedge c)$
$a \vee (b \wedge c) = (a \vee b) \wedge (a \vee c)$

basic logic (standard axioms)

$(a \barwedge a) \barwedge (a \barwedge a) = a$
$a \barwedge (b \barwedge (b \barwedge b)) = a \barwedge a$
$(a \barwedge (b \barwedge c)) \barwedge (a \barwedge (b \barwedge c)) =$ $((b \barwedge b) \barwedge a) \barwedge ((c \barwedge c) \barwedge a)$

basic logic (Sheffer axioms)

$((a \barwedge b) \barwedge c) \barwedge (a \barwedge ((a \barwedge c) \barwedge a)) = c$

basic logic (shortest axioms)

Fig. 34

So what that tells us is that the 50,000th axiom system that one finds in an enumeration of axiom systems is the axiom system for logic. That gives us a sense of where logic sits in the space of actual axiom systems, and of course one can then ask about all of the other axiom systems that weren't the 50,000th one that corresponds to logic, or the 30,000th one that applies to commutative group theory.

What about all these other axiom systems?

It turns out, as far as I can tell, there's nothing very special about the axiom systems that have been chosen to be studied by us. They're really something that are

very much based on an historical development of mathematics, and an interesting issue—what one can do with all these other axiom systems and what kinds of fields of mathematics can be developed—and I think that's an interesting direction for efforts in Theorema and other kinds of things, to try and see what else is out there in the space of mathematics.

I might just mention that this proof was done with Waldmeister which is based on an application of the unfailing completion algorithm that I guess is in some ways related to Bruno's work.

Let me end by saying that one of the things that's come out of the work that I've done is this idea that there's much more that could be in mathematics than has been so far investigated. I mean, when we look at these simple programs—rule 30, whatever else—these are processes, these are phenomena that have not traditionally been studied in mathematics, and when we look at the space of all possible simple programs, the space of all possible axiom systems, there's a lot else out there that hasn't been analyzed with existing mathematics, but it is something, as one automates existing mathematics one can expect to crack these generalizations of mathematics as well, and I hope and expect that will be a direction worth doing.

And I thought I might just mention *Mathematica*. Within *Mathematica*, you have sort of two different levels of operation. In one, the level that is very general for processing, that will apply to any processing of any of these possible systems. The other is this level of processing that corresponds to nineteenth-century polynomial operations and traditional computer algebra. One of the questions is whether there's an intermediate level that still captures the essence of what might be interesting in twentieth-century mathematics, but isn't as general as the arbitrary symbolic processing thing, but isn't less specific, as traditional nineteenth-century polynomial mathematics and so on.

But anyways, the main message I suppose of what I've done in natural sciences is that I've made quite a lot of progress in natural science by general sets of primitives that can be studied by what is traditional in mathematics and that idea can be applied back to mathematics as well.

There's a lot out there that could become fields of mathematics worth studying.

Let me stop there, thank you very much.

Towards a Symbolic Computational Philosophy (and Methodology!) for Mathematics

Doron Zeilberger

Dedicated to Bruno Buchberger, on the occasion of his 5!/2!-th Birthday

One of the most *profound* mathematical concepts is the **diagonal** (cf. Pythagors, Cantor, Gödel, Turing). The format of a diagonal element is *AA*, but *my* favorite element is *BB*. My favorite movie-star used to be *Brigit Bardot*, my favorite playwright is *Bertolt Brecht*, my favorite Sesame Street character is *Big Bird*, and my favorite *mathematician, computer scientist, logician, philosopher, pedagogue, administrator, clarinet-player*, and *human-being* is: **Bruno Buchberger** (henceforth **BB**).

Math and science are *emergent* phenomena, and a new field is created *spontaneously* by the efforts of many people. So it is over-simplistic to talk about the 'founder' of a field. But, at least by one *criterion*, **BB** founded **SC** (Symbolic Computation), since he created **JSC** (Journal of Symbolic Computation).

By the way, **SC=CA=CF**, where **CA** stands for *Computer Algebra*, and **CF** is the French name, *Calcul Formel*.

Definition of CA (c. 1980): "CA is the part of CS which designs, analyses, implements, and applies, algebraic algorithms".

This version: Feb. 20, 2003. First Version: Dec. 12, 2002. Edited version of the transcript of an invited talk given at the Bruno Buchberger Symposium, LMCS2002, RISC-Linz, Hagenberg castle, Hagenberg, Austria, Oct. 22, 2002, 12:15-13:05. Supported in part by the NSF.

D. Zeilberger (✉)
Department of Mathematics, Rutgers University (New Brunswick), Hill Center-Busch Campus, 110 Frelinghuysen Rd., Piscataway, NJ 08854-8019, USA
e-mail: zeilberg@math.rutgers.edu; http://www.math.rutgers.edu/~zeilberg/

This definition was given by Rüdiger Loos[7] in the seminal volume edited by BB, Collins and Loos[3]. As you can see, the beginning was rather modest, and CA only claimed to be a tiny part of computer science.

A better definition was given recently by **BB** himself [2].

Definition of SC (2002): "The part of math that can be expressed by quantifier-free predicate-logic is the **SC** part of math".

But, according to me, **SC** is a *primitive, fundamental* entity, and it is *math* that needs a definition. So here is my own 2050 definition of math.

Definition of Math (2050): **Math:=SC.**

Let's pause to talk about the first word of the phrase **Symbolic Computation**, i.e. on *Symbolic*. Symbolism is as old as humanity, and is all around us, in art, science, religion, and of course, language. Here is what Alfred North Whitehead had to say about *language*, in his delightful little book [9].

> "*Language is such a symbolism... The word is a symbol, and its **meaning** is constituted by the ideas, images, and emotions which it raises in the mind of the hearer...*".
> "*There is another sort of language, purely a written language, which is constituted by the mathematical symbols of the science of algebra*".

Whitehead then goes on to say that in *algebra* the *meaning* is irrelevant, since the *symbols do the reasoning for you*.

It is this *magical* property of algebra, that is today amplified million-fold by *computer algebra*, that makes *humans*, those *semantical* creatures, so ambivalent about it.

1 Bordeaux 1991

1991 was a very good vintage for combinatorics, since at that year the historic 3rd 'Formal Power Series and Algebraic Combinatorics' conference took place. It was historic because amongst the five invited speakers, one was a Bourbaki (Pierre Cartier), while another one was a software developer (Gaston Gonnet). The three other invited speakers were Gilbert Labelle, of species fame, Asymptotics Guru Phillipe Flajolet, and myself. In my talk I made the famous statement: **EXTREME UGLINESS IS BEAUTIFUL.**

It was made in defense of the proofs generated by my beloved electronic servant, Shalosh B. Ekhad, that to the uninitiated human look very unmotivated and ugly. I claimed that it was a new art form, and it is exactly their 'inelegance' that made them so elegant.

The above sentence falls under the *paradigm*: **Extreme X is the Opposite of X.**

This *symbolic sentence* gives rise, by specializing **X**, to many 'profound statements'. E.g. try: **X**=Simplicity, Kindness, Love, Hate, Modesty, Fame, ...

The *slogan* of the present talk is: **Extreme Abstraction is Concrete.**

And this is made possible by *algebra*, and especially by *computer algebra*.

The most salient feature of mathematics is its *abstraction*. This was made explicit in Tim Gowers's fascinating recent booklet [6], where he described the 'abstract method' and epitomized it by the slogan 'A mathematical object is what it does'.

2   A Short History of Abstraction

Once upon a time there were *three* bears, *three* lions, *three* apples. All of which were designated by three strokes on the cave's wall, and from this was born the very abstract *concept* 'three', denoted by the *symbol* 3.

Thus the statement $2 + 3 = 3 + 2$ is really a deep theorem, containing infinitely many facts: 'two bears and three bears is the same as three bears and two bears', 'two lions and three lions is the same as three lions and two lions', 'two apples and three apples is the same as three apples and two apples' , etc. Then humans discovered other such deep theorems: $4 + 7 = 7 + 4, 5 + 8 = 8 + 5, \dots$.

Much much later, came another leap. All these theorems were recognized as special cases of just *one*

Theorem. *Let a and b be arbitrary integers, then $a + b = b + a$.*

Here a and b are *symbols* that *symbolize* concrete numbers. As such this theorem requires a proof.

Proof. Since $a + 0 = 0 + a$, this is true for $b = 0$. Next let's prove this for $b = 1$. $a + 1 = ((a - 1) + 1) + 1 = (1 + (a - 1)) + 1 = 1 + ((a - 1) + 1) = 1 + a$. Now, $a + b = a + ((b - 1) + 1) = (a + (b - 1)) + 1$. By induction, this equals $((b - 1) + a) + 1$, which equals $(b - 1) + (a + 1) = (b - 1) + (1 + a) = b + (-1 + 1) + a = b + 0 + a = b + a$.

Traditionally, abstract symbols stand for more concrete objects, and they do have *meanings*.

In Symbolic Computation, a and b stand for *themselves*. In Maple, type(a, symbol); and type(b,symbol); are true. So $b + a := a + b$ by *fiat*, and it does not require proof. As our **Birthday Boy** said recently [2]:

> Math can be viewed as a network of meta-theories: A theorem on a meta-level may 'trivialize' the invention/proof of many theorems in the object level.

Every time we abstract from one level to the next meta-level we trade a **pound of Semantics for an ounce of Syntax**.

Here is a parable. The infinitely many theorems $(1 + 2)^2 = 1^2 + 2 * 1 * 2 + 2^2$, $(2 + 3)^2 = 2^2 + 2 * 2 * 3 + 3^2$, $\dots$, can be abstracted to the general 'theorem' $(a + b)^2 = a^2 + 2ab + b^2$, that if you think of a and b as symbols standing for numbers, requires a proof using the *axioms* of algebra. But if you encapsulate these 'axioms' into combinatorial rewriting rules (expand in Maple), and since there is a *canonical form* algorithm, this is now a purely routine fact, of the same epistemological stature as $2 + 2 = 4$.

In the same vein, $(a + b)^3 = a^3 + 3a^2b + 3ab^2 + b^3$ is purely routine, as well as the binomial theorem for $(a + b)^n$, for any specific n, although these theorems get *deeper and deeper* (i.e. requiring more time and memory) as n gets bigger, and if you use expand to prove it for $n = 10^{100}$, Maple will run out of time and memory.

Until 1988, the **Binomial Theorem**

$$(a+b)^n = \sum_{k=0}^{n} \binom{n}{k} a^k b^{n-k} \quad,$$

for arbitrary n, was considered a genuine *theorem*. But thanks to **Zeilberger's algorithm**, which is part of **WZ theory**, it is now completely routine, since both sides have the *canonical form* $N - (a + b), 1$. Here I used the shorthand for describing so-called holonomic sequences by giving the operator annihilating them (where N is the shift operator), followed by the "initial conditions".

Similarly, using the multisum procedure, the *trinomial* theorem, the *quartonomial theorem*, etc. are all routine, as is the *multinomial theorem* for each fixed number of variables. But we would have to wait for **WZ theory, Chap. II**, to make the full *multinomial theorem*

$$\left(\sum_{i=1}^{k} a_i \right)^n = \sum_{i_1 + \cdots + i_k = n} \frac{n!}{i_1! \ldots i_k!} a_1^{i_1} \cdots a_k^{i_k}$$

with *symbolic k*, fully routine.

Not only **Math** progresses from the *concrete* to the *abstract*, also **Music** (e.g. Schoenberg, Stravinsky), **Art** (e.g. Picasso), **Literature** (e.g. Proust, Joyce) and even **Religion**.

The old testament God, Yehova (Jehovah), who was already much more abstract than His idolic predecessors, still enjoyed the good life, savoring the odor that came from the animal sacrifices offered by the *cohanim*. In comparison, the loving God of the new testament is much more civilized and abstract. Also Jewish Law progressed towards symbolism and abstraction. For example, the barbaric *eye for an eye* and *tooth for a tooth*, was changed, by Rabbi Hillel, to paying fines, simply by defining certain units of money to be called *eye* and *tooth*, that is replacing a real eye by a symbolic eye.

Yet greater feats of abstraction were achieved by the medieval Rabbi Abraham Ibn Ezra (alias Ben-Ezra), and, inspired by the latter, at the dawn of the Age of Enlightenment, by Baruch Spinoza. Sometimes being too abstract may risk your life, or at least make your life miserable. Spinoza was ex-communicated from the Jewish community of Amsterdam for his 'heresy'.

This 'abstraction trick' can be a very powerful argument in giving 'modern' humans *reasons to believe*. Here is a quotation from a great contemporary Christian theologian.

> The resurrection of Jesus was **not** a historical event, but it is **symbolic**, affirming that death does not have the **last** word about human life.

The above quotation is taken from Rev. Maurice Wiles's very interesting little book entitled *Reason to Believe*[10]. I don't think that it is a coincidence that the child of a great theologian turned out to be a great mathematician, since both math and religion are based on *abstract concepts*.

3   The Rise and Fall of Logocentrism

Alfred North Whitehead claimed that all western philosophy is a footnote to Plato. The analogous thing may be said about Euclid and (western) mathematics. Indeed, the Euclidean axiomatic approach dominated mathematics for the last 2,000 years and was considered a paradigm of rigor, that philosophy and other branches of knowledge tried to emulate. For example, Spinoza used the Euclidean mold of axioms and propositions to write his Ethics book. Also modern economics, in its struggle to become 'scientific', often uses axioms.

The precise logical thinking in Euclid's *elements* was also considered to be of great pedagogical, moral, and even *religious* significance, since the 'rational' thinking that studying it was supposed to engender supposedly made one a better person.

Ironically, the greatest *rationalist* of them all, René Descartes, demolished the Euclidean supremacy, at least in its immediate, geometrical, scope, by creating *Analytic Geometry*. I will soon argue that Descartes's breakthrough also shattered the Euclidean tradition in the *general* sense, in putting *Algebra* above *Logic*. However this, more general sense, took more than 300 years to kick in.

Even though René Descartes already trivialized Geometry *in principle*, it took computer algebra, and the genius of our birthday boy, Bruno Buchberger, to make this *trivialization*, or more politely, *algorithmization*, feasible in *real time*.

What Buchberger did, via his revolutionary Gröbner bases, was to establish a *canonical form* for ideals in a polynomial ring. Since every entity in Plane Geometry can be described by an ideal, it gives us a decision procedure for proving any theorem. So one no longer must be clever, or have geometrical intuition. All that is needed is a knowledge of typing, the rest is done by the computer.

Alas, the strength of Gröbner bases is also their weakness. Because they can do so much, they are usually very slow, and one has the exponential-time-and-space curse. It turns out that for most theorems in Plane (and space) Geometry, one does not need ideals, and one can still prove things by making everything explicit, since via *parameterization*, things run much faster. Here is an example of Erdös's favorite theorem, called the **Butterfly Theorem**. Its statement, in Maple, is:

```
Butterfly:=proc() local P,t,i,R,Li,M,X,Y:for i from 1
to 4 do
P[i]:=ParamCircle([0,0],R,t[i]) od:M:=Pt(Le(P[1],P[3]),
Le(P[2],P[4])):
Li:=PerpPQ([0,0],M):X:=Pt(Le(P[1],P[4]),Li):Y:=Pt(Le
(P[2],P[3]),Li):
ItIsZero(DeSq(M,X)-DeSq(M,Y)):end:
```

This is taken from Shalosh B. Ekhad's **Geometry Textbook**[5]. It is a textbook written entirely in Maple, but that is fun to read for computers and humans alike. The above is a complete statement of the Butterfly theorem, and in order to prove

it, all one has to do is type `Butterfly();`. In less than a second, the computer returns `true`.

For the sake of completeness, here are those macros that are needed for `Butterfly`. So together with these, we have a completely self-contained *statement* and *proof* of the theorem.

```
#Def (Area of triangle ABC)
AREA:=proc(A,B,C):normal(expand(((B[1]*C[2]-B[2]*C[1]-
A[1]*C[2]+A[2]*C[1]-B[1]*A[2]+B[2]*A[1])/2)):end:
```

```
#Def (Square of the distance of points A and B)
DeSq:=proc(A,B):(A[1]-B[1])**2+(A[2]-B[2])**2: end:
```

```
#Def (Is it zero?)
ItIsZero:=proc(a):evalb(normal(a)=0):end:
```

```
#Def (The eq. of the line joining A and B)
Le:=proc(A,B) AREA(A,B,[x,y]):end:
```

```
#Def (Generic point on a parametric circle center [c[1],
c[2]] and radius R)
ParamCircle:=proc(c,R,t):[c[1]+R*(t+1/t)/2,c[2]+R*(t-1/
t)/2/I]:end:
```

```
#Def (Line through Q perpendicular to PQ)
PerpPQ:=proc(P,Q):expand(((y-Q[2])*(P[2]-Q[2])+(x-Q[1])*
(P[1]-Q[1]))):end:
```

```
#Def (The point of intersection of lines Le1 and Le2)
Pt:=proc(Le1,Le2) local q:q:=solve(
numer(normal(Le1)),numer(normal(Le2)),x,y):
[normal(simplify(subs(q,x))),normal(simplify(subs(q,y))
)]:end:
```

The reason that it took less than 1 s of CPU time was that I used the *parametric equation* of the circle (implemented in `ParamCircle` above). In Humanese it is

$$x = R(t + t^{-1})/2 \quad , \quad y = R(t - t^{-1})/(2i).$$

Notice that everything is pure (high-school) algebra.

Now, had we insisted on using Gröbner bases, then, in the 'straightforward approach' we would have had to write the assumptions that the four points $P_1 = (x_1, y_1)$, $P_2 = (x_2, y_2)$, $P_3 = (x_3, y_3)$, $P_4 = (x_4, y_4)$ lie on the circle, by introducing eight variables $x_1, y_1, x_2, y_2, x_3, y_3, x_4, y_4$, and introducing the ideal generated by $\{x_1^2 + y_1^2 - 1, x_2^2 + y_2^2 - 1, x_3^2 + y_3^2 - 1, x_4^2 + y_4^2 - 1\}$. (Dongming Wang pointed out that it is much more efficient to take three points, find their circumcircle,

and demand that the fourth point lies on that circle). Next the computer would have had to compute its Gröbner basis. Then compute the *bottom-line* quantity (that has to be shown equal to 0) and finally use `normalf` to find it modulo the above ideal. This takes a little longer than before.

Of course, as you all know, Gröbner bases can do many other things besides proving 'high-school geometry' theorems, and it is impossible to imagine modern (and post-modern) algebra without it.

In all the many theorems in Ekhad's Geometry text, it was possible to avoid using Gröbner bases. Of course for *general* problems, one can't avoid it, but it so happens that for most theorems that come up in real life, it was possible to get away with parameterization (essentially equivalent to using Trigonometry). Here we have an example of a **targeted** subansatz. (Dongming Wang pointed out that the many cases where one can do without Gröbner bases are those where it is possible to 'linearize' quadratic relations.)

By successive abstraction, and going from one level to the next meta-level, one gets more and more symbolic. But once the symbols can be considered **qua symbols**, regardless of their *meaning*, things become **concrete** again. Ideally one should be able to write a computer program, and there is nothing more concrete than a computer program. So indeed **Ultimate Abstraction** is **Concrete**.

Analogously, **Ultimate Semantics is Syntactic**. But the converse is also true: **Ultimate Syntax is Semantic**. The symbols themselves are *marks on the paper* (so despised by Brouwer), and today *bytes*, obeying certain *combinatorial* conditions. So everything boils down to *combinatorics*, and what can be more concrete than combinatorial, finite, objects?

The **Axiomatic Method**, and **Formal Logic** (starting with Euclid and Aristotles) have had their 2,000 years of glory. Thanks to Kurt Gödel we know that they can't do *everything* (even in principle). Another giant, Gregory Chaitin, quantified it with his beautiful uncomputable constant Ω. Gödel also proved that even for decidable statements, there exist short theorems with very long proofs. The Four-Color Theorem may be an example.

Hence, **humans** (and even **computers**) can formally and fully rigorously only prove facts of very low complexity (in the technical sense of computational (or program-length) complexity).

Realizing this, if we want to transcend our intrinsic triviality we need to **diversify**, and be more **inclusive**. We should welcome the whole gamut of mathematical truths. In addition to fully rigorous proofs, we also need semi-rigorous proofs (see [13]), ϵ-rigorous proofs, non-rigorous proofs, very plausible conjectures, plausible conjectures, all the way to wild guesses. Of course, for interesting statements it would be nice to upgrade their level of truth, but we should not spend too much time on these upgrades, since there are so many exciting new facts to discover.

We desperately need a new philosophy and methodology for doing mathematics, and I believe that the *practice* of symbolic computation is a very good *rough draft*, and *starting point* for this.

But before describing it, let's digress for 1 min in order to critique the prevailing philosophies.

Formalism is too logocentric, while **Logicism** is even more logocentric. **Intuitionism** is too human-centric while **Humanism** (started by Phillip Davis and Reuben Hersh and fully developed by Hersh) is even more human-centric. **Platonism** is too platonic and metaphysical, while **Bourbakism** is too structured and semantical.

The **"new" philosophy** and **methodology** that I am proposing here, inspired by computer algebra, is really a revival of two very old traditions:

Pythagorianism, with its denial of 'real' numbers and the infinite (see [14]).

The Hindu-China-Babylonian-Persian-Arabic tradition of 'high-school', algorithmic, algebra.

In the new **methodology**, we should forget both about **syntax** and **semantics**, and focus instead on **PRAGMATICS**!: Learn to serve IRH (Its Royal Highness), the **Computer**.

We also need **Problem-Solving** methodologies, e.g. adapt Polya's heuristics (as given in his famous classic book 'How to Solve It' and the series of books on plausible reasoning), to Computers and especially Computer Algebra Systems.

A very **important Principle**, originally intended as a **pedagogical principle**, but that I am sure has a much wider **scope** for doing **research** is, The **BBBBwBp**, which is short for: Bruno Buchberger's Black Box White Box Principle.

This great brainchild of our beloved Birthday Boy [1] means that when we teach students a new concept or method or algorithm, in the first phase we should **not** let them use computers, but let them do it by hand, so that they can internalize what they are learning, with simple examples. But once they mastered it, it should be encapsulated into a **black box**, so that they can graduate to bigger, better, and deeper things, without being bogged down with details. This principle should be used recursively, of course, until an intricate web of knowledge will mature in the student's head.

But all of us are **students**! The computer is our **master**, and if we will learn to use **black boxes** efficiently, trusting their contents, without necessarily fully understanding them, we will be able to go much further. One of the reason mathematicians made so little progress so far is their obsession for knowing all the details, and not trusting previous results as black boxes. Furthermore, because math is so fragmented and specialized, there are lots of black boxes, developed by specialists in other areas, that are unaccessible to us, because of the language-barrier between sub-specialties. We really need an **Esperanto**, or at least a **lingua franca** that will bridge this tragic communication failures between mathematical subareas. A good start could be Maple (or Mathematica), that of course will have to keep expanding. The language of formal logic, with its quantifiers, turned out to do more harm than good because of its overwhelming generality and pedantry, and because it did not describe the day-to-day practice of doing mathematics. (I know that Bruno may disagree with me here, since he is a great fan of logic, and I agree that logic is a marvelous thing, but it should not be carried too far.)

The Bruno Buchberger Black Box-White Box principle is a great example of using wisdom gained in **teaching** to help do **research**. Most research mathematicians either dislike teaching, viewing it as a chore, or like it, but think of it as an activity unrelated to their research. This 'binary-opposites' pair, *teaching-research*, is yet another dichotomy that has to be abandoned. In the future research will be both **teaching** and **learning**. First teaching the computer to do mathematical research, and later learning from its output, and so on indefinitely.

One possible way to teach computers is to look for new **Ansatzes**, that will convert classes of theorems to 'high school algebra', amenable to computer search. I call this approach the **Ansatz Ansatz**, in analogy to Thomas Kuhn's approach to the philosophy of science, that may be called the **Paradigm Paradigm**.

Sometimes you have to transcend to a **superansatz**. Let's make-believe that the sequence $f(n)$ enumerating the number of legal bracketings with n pairs of brackets is 'intractable' (it is in fact the sequence of Catalan numbers, that satisfy a *non-linear* recurrence with constant coefficients, and a linear recurrence with *polynomial* coefficients, but let's forget this right now and pretend that we are linear, constant-coefficients, creatures). If you consider instead the more *general* function $F(m,n)$ of *prefixes* of legal bracketings, with m left parentheses and n right parentheses (and of course $m \geq n$), then $F(m,n)$ satisfies a *partial recurrence equation* $F(m,n) = F(m,n-1) + F(m,n-1)$ with the boundary conditions $F(m,0) = 1$ and $F(m,m+1) = 0$, from which follows immediately that $F(m,n) = (m-n+1)(m+n)!/((m+1)!n!)$ (check!), and in particular, $f(n) = F(n,n) = (2n)!/((n+1)!n!)$.

What happened here was that we went up to a **superansatz** that made the problem more tractable.

Sometimes the ansatz is adequate, *in principle*, but is computationally inefficient. In that case one can look for **targeted subansatzes** that work faster for subclasses of problems or objects. We already saw that in the Plane Geometry example above, where Gröbner bases are relatively fast, but can be made yet faster with the 'right' parametrization, for an important subclass, by staying in the *rational function ansatz*.

Another example is the **Holonomic paradigm**, that plays a fundamental role with respect to algorithmic work in the field of special functions (e.g. NIST's Digital Mathematical Functions Library). My slow algorithm [12] (vastly improved by Chyzak[4], and in fact it is not so slow anymore at all), is very general. Then my 'fast' algorithm [11] is much faster, but can only do *proper hypergeometric* summands. Finally the WZ-pairs are yet faster, and also very elegant, but only work when the right hand side is closed form (i.e. the recurrence outputted by Zeilberger's algorithm is first-order).

Another important subansatz is Wegschaider's use of ideas of Verbaeten, that can be seen as a targeted subansatz for WZ/Sister Celine in the multiple sum case. Recently there was an exciting speed-up achieved by Axel Riese and Burkhard

Zimmermann obtained by supplementing this with another paradigm, namely with random modular checking.

Peter Paule's recent exciting work on *contiguous relations* can be viewed as a *superansatz* of my creative telescoping ansatz and is analogous to the above-mentioned generalization of the Catalan function f(n) to its bivariate version F(m,n).

Once you understand your **ansatz**, you can have your computer, discover **from scratch**, all the theorems in the field up to any given complexity.

4 A Confession

The masterpiece [5] that was supposedly 'downloaded from the future' was entirely written, by hand, by a human (myself). All that (the current) Shalosh B. EKhad (IV) did was to *run the program* and *prove correctness* for all the 'statement-proofs', and also draw the beautiful diagrams.

Not only did I lie about the author and date, but I did something much worse. In the 'Preface by the downloader', I pretended that the text was automatically generated by the computer Shalosh B. Ekhad, XIV, by using another *meta-program* that started with three generic points, and then by iterating a few macros, viz. **Pt, Le, Ce**, the computer kept getting new points, lines, circles, etc. Whenever a new object coincides with a previously defined one, the computer *discovered*, and at the very same time *proved*, a new theorem,

But, even though it was **not** done that way, it **could have been done that way**, I was just too lazy. I hope that someone will soon write this 'Geometry from scratch' computer program.

5 Automated DISCOVERY (and Proof) of ALL Binomial Coefficient Identities (Up to a Prescribed Complexity)

The so-called Zeilberger algorithm can *prove* any binomial coefficient identity (alias hypergeometric series) identity, once conjectured. WZ theory can do even better. By starting with one of the **golden oldies**, like Gauss, Saalschutz, Dixon, or Dougall, and *specializing and dualizing* (see [8]), it can discover, and automatically prove, lots of *new* 'strange' hypergeometric identities.

BUT suppose that you do not know anything about the human heritage. Do you really need these humans to get started? Of course you don't! In principle all that humans have to do is *define* a concept. In this case the appropriate concept is that of *WZ pair*, invented by the humans Herb Wilf and Doron Zeilberger.

Let's first review *WZ theory*.

6   WZ Theory in a Nutshell

Suppose you want to prove an identity of the form

$$\sum_k NICE(n,k) = NICE'(n) \quad ,$$

where 'NICE' means hypergeometric in its arguments, i.e. $(NICE(n+1,k)/NICE$ (n,k) and $NICE(n,k+1)/NICE(n,k)$ are rational functions of (n,k) and $NICE'(n+1)/NICE'(n)$ is a rational function of n). Actually, we need $NICE(n,k)$ to be *proper-hypergeometric* (see [8]), but let's forget about this technicality now.

The first step is to divide by $NICE'(n)$, and since $NICE(n,k)/NICE'(n)$ is also nice, let's rename the latter $F(n,k)$ and try to prove

$$\sum_k F(n,k) = 1 \quad . \qquad\qquad (NiceIdentity)$$

It so happens that in 99 % of the cases.[1] The *raison d'être* for such an identity is the existence of another *nice* discrete function of (n,k), let's call it $G(n,k)$, the so-called **WZ-mate**, such that

$$F(n+1,k) - F(n,k) = G(n,k+1) - G(n,k) \qquad\qquad (WZ)$$

Furthermore, $G(n,k)/F(n,k)$ is a *rational function* $R(n,k)$, called the *certificate*. Once $R(n,k)$ is given it is routine to prove (WZ), since defining the rational functions

$$R_1(n,k) = \frac{F(n+1,k)}{F(n,k)} \quad , \quad R_2(n,k) = \frac{F(n,k+1)}{F(n,k)}. \qquad (Ratios)$$

and dividing (WZ) by $F(n,k)$, reduces the proof of (WZ), at any given case, to the verification of the routine identity amongst rational functions (and by clearing denominators, amongst polynomials in n, k):

$$R_1(n,k) - 1 = R(n,k+1)R_2(n,k) - R(n,k) \quad . \qquad (WZ')$$

Now once (WZ') is established, (NiceIdentity) follows immediately, by summing (WZ) with respect to k, yielding that $a(n) := \sum_k F(n,k)$ satisfies $a(n+1) - a(n) = 0$, and hence that $a(n)$ is a constant, and to prove that constant is 1 all we have to do is verify that $a(0) = 1$.

---

[1] There is always something more general that guarantees a proof.

Of course the "we" above is done completely automatically by the computer, as is the discovery of $R(n, k)$ (whenever it exists), which is done by applying Gosper's algorithm (w.r.t. k) to $F(n + 1, k) - F(n, k) = (R_1(n, k) - 1)F(n, k)$.

But, how can we find *new* identities, *completely from scratch*? Every WZ-pair is really a *miracle*. First note that the rational functions $R_1(n, k)$ and $R_2(n, k)$, in order to arise from a hypergeometric $F(n, k)$ by (*Ratios*), must satisfy the obvious compatibility condition

$$R_1(n, k + 1)R_2(n, k) = R_1(n, k)R_2(n + 1, k). \qquad (Compatibility)$$

So the (*WZ*) miracle hinges on the existence of a triple of rational functions (R_1, R_2, R) depending on (n, k) such that the non-linear equations (*Compatibility*) and (*WZ'*) are satisfied.

However, not all such triples are *interesting*, since for any nice $A(n, k)$, defining

$$F(n, k) = A(n, k + 1) - A(n, k) \quad , \quad G(n, k) = A(n + 1, k) - A(n, k),$$

automatically yields a (*WZ*)-pair, from which we can get lots of trivial solutions for the system (*Compatibility*), (*WZ'*). In order to get the 'nice and interesting identities' we must 'mod out' by these 'exact forms'.

7 Conclusion

The SymbolicComputational tail is starting to wag the mathematics dog, and will continue to do so more and more vigorously, and very soon, (traditional) math will become the tail of computer algebra. In order to accomplish this rôle-reversal, we desperately need a new outlook, and **working style**, and this article proposes a very rough draft. The details of the emerging philosophies and methodologies are yet to be worked out, but one thing is certain: the **pioneers** of the **SymbolicComputational Revolution**, most notably **Bruno Buchberger**, will become legendary and mythical heroes as long as human beings (and/or computers) will continue to do mathematics.

Acknowledgements I wish to thank Bruno Buchberger and Dongming Wang for numerous suggestions and comments on an earlier draft.

References

1. Buchberger, B.: Should students learn integration rules. SIGSAM Bull. **24/1**, 10–17 (1990)
2. Buchberger, B.: Theorema and mathematical knowledge management. Lecture delivered at the IMA (Minneapolis), Aug 1, 2002, available from http://www.risc.uni-linz.ac.at/people/buchberg/invited_talks.html

3. Buchberger, B., Collins, G.E., Loos, R.: Computer Algebra Symbolic and Algebraic Computation, 2nd edn. Springer, New York (1982)
4. Chyzak, F.: An extension of Zeilberger's fast algorithm to general holonomic functions. Discret. Math. **217**, 115–134 (2000). Preliminary version available on-line at http://algo_inria_fr/chyzak/publications.ps
5. Ekhad, S.B. XIV:Plane Geometry: An Elementary School Textbook (c. 2050), (2005) http://www.math.rutgers.edu/~zeilberg/GT.html
6. Gowers, T.: Mathematics – A Very Short Introduction. Oxford University Press, London (2002)
7. Loos, R.: Introduction to [3], pp. 1–11
8. Petkovsek, M., Wilf, H., Zeilberger, D.: A=B. A. K. Peters, Wellesley (1996). [On-line version available from the authors' websites]
9. Whitehead, A.N.: Symbolism. Macmillan, New York (1927). (Reprinted by Fordham University Press, 1985)
10. Wiles, M.: Reason to Believe. Trinity Press International, Harnshurg (1999)
11. Zeilberger, D.: A fast algorithm for proving terminating hypergeometric identities. Discret. Math. **80**, 207–211 (1990). Available on-line from http://www.math.rutgers.edu/~zeilberg/mamarim/mamarimhtml/fast.html
12. Zeilberger, D.: A Holonomic systems approach to special functions identities. J. Comput. Appl. Math. **32**, 321–368 (1990). Available on-line from http://www.math.rutgers.edu/~zeilberg/mamarim/mamarimhtml/holonomic.html
13. Zeilberger, D.: Theorems for a price: tomorrow's semi-rigorous mathematical culture. Notices Am. Math. Soc. **40**(8), 978–981 (1993). (Reprinted in Math. Intell. **16**(4), 11–14 (Fall 1994))
14. Zeilberger, D.: "Real" analysis is a degenerate case of discrete analysis. In: Aulbach, B. (ed.) Proceedings of ICDEA7. Taylor and Francis (2001, to appear), (2004). Available on-line from http://www.math.rutgers.edu/~zeilberg/mamarim/mamarimhtml/real.html

Printed in the United States
By Bookmasters